FATIGUE of MATERIALS II
ADVANCES and EMERGENCES in UNDERSTANDING

Cover Photograph: Scanning electron micrographs of the fatigue fracture surface of the 9310 sample cyclically deformed at load ratio of -1 at a maximum stress of 467.8 MPa, resultant fatigue life of 385,763 cycles, showing well defined, yet shallow striations, in the region of stable crack growth.

(from K. Manigandan, T.S. Srivatsan, T. Quick and A. M. Freborg)

FATIGUE of MATERIALS II
ADVANCES and EMERGENCES in UNDERSTANDING

Proceedings of a symposium sponsored by

Mechanical Behavior Committee of
The Minerals, Metals & Materials Society (TMS) and
ASM International

held during

Materials Science & Technology 2012 (MS&T'12)
October 7-11, 2012 in Pittsburgh, Pennsylvania

MS&T'12®

Edited by

T.S. Srivatsan **M. Ashraf Imam**

R. Srinivasan

Editors
T.S. Srivatsan
M. Ashraf Imam
R. Srinivasan

ISBN 978-3-319-48583-6 ISBN 978-3-319-48105-0 (eBook)
DOI 10.1007/978-3-319-48105-0

Chemistry and Materials Science: Professional

Published by Springer International Publishers, Switzerland, 2016
Reprint of the original edition published by John Wiley & Sons, Inc., 2013, 978-1-11852-093-2

Printed on acid-free paper

This Springer imprint is published by Springer Nature
The registered company is Springer International Publishing AG
The registered company address is: Gewerbestrasse 11, 6330 Cham, Switzerland

TABLE OF CONTENTS
Fatigue of Materials II: Advances and Emergences in Understanding

Symposium on the Fatigue of Materials II: Advances and Emergences in Understanding

Overview I

Overview II

Advanced Materials I

Advanced Materials II

Other Materials

PREFACE

This bound volume contains the papers presented in the second international symposium entitled *"Fatigue of Materials: Advanced and Emergences in Understanding"* held during the 2012 Materials Science and Technology meeting at Pittsburgh, PA, USA in October 2012. The five-session symposium was sponsored by the Mechanical Behavior of Materials Committee of ASM International and the Minerals, Metals and Materials Society (TMS). It is essentially the sixth in a series of symposia on the subject topic *"Fatigue of Materials"*. The first three symposia focused on *"Deformation and Fracture of Ordered Intermetallic Materials"*. The first symposium was held in Pittsburgh (PA, USA) in 1993; the second symposium was held in Rosemont (Illinois, USA), in 1994, and the third symposium was held in Cincinnati (Ohio, USA) in 1996. The fourth symposium was in honor of Professor Paul C. Paris and focused on *"High Cycle Fatigue of Structural Materials"* and held in Indianapolis (Indiana, USA) in 1997. The fifth symposium related to *Advances and Emergences in Understanding* was held in Houston, Texas in October 2010

This symposium, the sixth in a series on the topic of "Fatigue of Materials", was well represented with abstracts from engineers, technologists and scientists from academia, research laboratories and industries, located both within the USA and few countries overseas. The forty plus abstracts that were approved for presentation at the symposium were divided into five sessions:

(a) Session 1: Overview I
(b) Session 2: Overview II
(c) Session 3: Advanced Materials I
(d) Session 4: Advanced Materials II
(e) Session 5: Other Materials

The abstracts chosen for presentation cover a broad spectrum of topics that represent the truly diverse nature of the subject that has grown both in strength and stature to become a key area of scientific and applied research. We have made every attempt to bring together individuals who could in a positive way put forth the advances while concurrently striving to enhance our prevailing understanding of the topic of Fatigue of Materials. We extend our warmest thanks and appreciation to the authors and session chairmen for their enthusiastic commitment and participation.

We also extend our most sincere thanks and appreciation to elected representatives that serve on the Mechanical Behavior of Materials Committee of ASM Int. and TMS for their understanding and acknowledgement of our interest, and approving our request to organize this intellectually stimulating event, i.e., second international symposium. An overdose of special thanks, gratitude and appreciation are reserved and extended to **Ms. Trudi Dunlap** (Programming Manager at The Minerals, Metals and Materials Society

(TMS) for her sustained attention, assistance, interest, involvement and participation stemming from understanding. This enabled in ensuring a timely execution of the numerous intricacies related to smooth orchestration and layout of this symposium from the moment immediately following its approval and up until compilation and publication of this proceeding (bound volume). At moments of need, we the symposium organizers have found her presence in TMS and participation to be a pillar of support, courteous, understanding, professional and almost always enthusiastically helpful and receptive. Timely thanks and assured appreciation is also extended to **Mr. Robert Demmler** (Graphic Designer, TMS) for his efforts and enthusiasm in ensuring organization and compilation of the material in a cohesive and compelling manner. The timely compilation and publication of this bound volume would not have been possible without the cooperation of the authors and the conference/symposium publishing staff at TMS (Warrendale, PA, USA).

We truly hope that this bound volume will provide engineers, scientist and technologists with new perspectives and directions in their research endeavors towards evaluating, understanding and improving the fatigue behavior of materials, spanning the entire spectrum to include both engineering materials and engineered materials.

Dr. T. S. Srivatsan
Division of Materials Science and Engineering
Department of Mechanical Engineering
The University of Akron
Akron, Ohio 44325, USA
Phone: 330-972-6196
E-Mail: tsrivatsan@uakron.edu

Dr. M. Ashraf Imam
Naval Research Laboratory
Materials Science & Technology Division
Naval Research Laboratory
4555, Overlook Avenue, SW
Washington DC, 20375, USA
Phone: 202-767-2185
E-Mail: ashraf.imam@nrl.navy.mil

Dr. R. Srinivasan
Mechanical and Materials Engineering Department
Wright State University
3640 Col. Glenn Highway
Dayton, Ohio 45435 USA
Phone: 937-775-5093;
E-Mail: raghavan.srinivasan@wright.edu

SYMPOSIUM ORGANIZERS

Dr. T.S. Srivatsan, Professor of Materials Science and Engineering in the Department of Mechanical Engineering at The University of Akron. He received his graduate degrees [Master of Science in Aerospace Engineering (M.S. 1981) and Doctor of Philosophy in Mechanical Engineering (Ph.D. 1984)] from Georgia Institute of Technology. Dr. Srivatsan joined the faculty in The Department of Mechanical Engineering at The University of Akron in August 1987. Since joining, he has instructed undergraduate and graduate courses in the areas of Advanced Materials and Manufacturing Processes, Mechanical Behavior of Materials, Fatigue of Engineering Materials and Structures, Fracture Mechanics, Introduction to Materials Science and Engineering, Mechanical Measurements, Design of Mechanical Systems and Mechanical Engineering Laboratory. His research areas currently span the fatigue and fracture behavior of advanced materials to include monolithic(s), intermetallic, Nano-materials and metal-matrix composites; processing techniques for advanced materials and nanostructure materials; inter-relationship between processing and mechanical behavior; electron microscopy; failure analysis; and mechanical design. His funding comes primarily from both industries and government and is of the order of a few millions of dollars since 1987. A synergism of his efforts has helped in many ways to advancing the science, engineering and technological applications of materials.

He has authored/edited/co-edited forty-nine (49) books in areas cross-pollinating mechanical design; processing and fabrication of advanced materials; deformation, fatigue and fracture of ordered intermetallic materials; machining of composites; failure analysis; and technology of rapid solidification processing of materials. He serves as co-editor of *International Journal on Materials and Manufacturing Processes* and on the editorial advisory board of few journals in the domain of Materials Science and Engineering. His research has enabled him to deliver over one-hundred ninety (190) technical presentations in National and International meetings and symposia; Technical/Professional societies; and Research and Educational institutions. He has authored and co-authored over six-hundred plus (600+) archival publications in International Journals, Chapters in books, Proceedings of national and international conferences, Reviews of books, and Technical Reports. In recognition of his efforts, contributions and its impact on furthering science, technology and education he has been elected Fellow of the society of *American Society for Materials, International (ASM Int.)*; Fellow of *American Society of Mechanical Engineers (ASME)*; and Fellow of *American*

Association for the Advancement of Science (AAAS). He has also been recognized as Outstanding Young Alumnus of Georgia Institute of Technology, and outstanding Research Faculty, the College of Engineering at The University of Akron. He offers his knowledge in research services to the U.S. Government (U.S. Air Force and U.S. Navy), National Research Laboratories, and industries related to aerospace, automotive, power-generation, leisure-related products, and applied medical sciences. He has the distinct honor of being chosen, in recent years, for inclusion in the following

(i) Who's Who in American Education;

(ii) Who's Who in the Midwest;

(iii) Who's Who in Technology;

(iv) Who's Who in the World;

(v) Who's Who in America;

(vi) Who's Who in Science and Engineering,

(vii) Who's Who among America's Teachers, and

(viii) Who's Who among Executives and Professionals (Cambridge).

Dr. M. Ashraf Imam is a Research Metallurgist at Naval Research Laboratory (NRL) pursuing basic research on material structure-property relationship. He also holds the position of Adjunct Professor of Materials Science at George Washington University in Washington, DC. Dr. Imam obtained a D.Sc. degree in Materials Science from George Washington University and an M.S. from Carnegie-Mellon University in the field of Metallurgy and Materials Science. Dr. Imam has demonstrated a unique ability to analyze the technical problems, and designing various experimental and theoretical techniques to arrive at optimum solutions. Often these investigations have resulted in technological breakthroughs leading to useful and unique applications. His publications (over 200) in a variety of refereed journals ranging from *Acta Metallurgica* to Philosophical Magazine letters are a clear testimony of Dr. Imam as a researcher and teacher with outstanding creativity and resourcefulness in the area of Materials Science and Engineering. His papers cover a variety of topics with a unifying theme of microstructure, characterization and analysis. His expertise on titanium resulted in editing three books on titanium including a large chapter on titanium in *Kirk-Othmer Encyclopedia of Chemical Technology*. These contributions to the literature have been widely acclaimed and recognized. Dr. Imam is FELLOW of *American Society for Materials International (ASM Int.)*. He is also a recipient of George Kimbell Burgess Award. The Award is given in recognition of outstanding achievement in research or administration to a member of the ASM Washington, DC chapter who has made original contributions in the field of metallurgy, materials, or mechanics or who demonstrates outstanding leadership in those fields. This award is given by the Washington, DC chapter of the *American Society for Materials (ASM-International).*

Dr. Raghavan Srinivasan, Professor in the Mechanical and Materials Engineering Department, is the *Director of the Materials Science and Engineering Program* at Wright State University, Dayton, Ohio. He received his PhD from the State University of New York, Stony Brook, and his MS from the University of Florida, Gainesville, both in Materials Science and Engineering, after completing his bachelors degree from the Indian Institute of Technology, Madras. He has over 100 technical publications in the areas of processing and mechanical behavior of advanced materials. He has edited two books, and contributed to the ASM Handbook series. His work has resulted in 4 patents and three patent applications which are under review. Dr. Srinivasan is a member of Tau Beta Pi, Alpha Sigma Mu, and Phi Kappa Phi honor societies, and has been recognized as Fellow of Alpha Sigma Mu, the International Honor Society for Materials Science and Engineering. He is a member of TMS, ASM-International, and SAMPE. He serves as an evaluator for ABET, and is a member of several committees, including the Accreditation, Professional Registration, Mechanical Behavior, and Shaping and Forming Committees of TMS, and is an associate member of the Failure Analysis Committee of ASM-International. Dr. Srinivasan is a registered Professional Engineer in the State of Ohio.

SESSION CHAIRS

Dr. René Alderliesten
Delft University of Technology (TU Delft)
The Netherlands

Dr. Nikhilesh Chawla
Arizona State University
Tempe, Arizona 85287, USA

Dr. R. Ebara
Fukuoka University
Fukuoka City, Japan 814-0180

Dr. Tarun Goswami
Wright State University
Dayton, Ohio 45435, USA.

Dr. H. Kang
The University of Michigan
Dearborn, Michigan, USA

Dr. Chandra S. Pande
Naval Research Laboratory
Washington DC, 20375, USA

Dr. N.Ranganathan
Laboratoire de Mécanique et Rhéologie
Polytech Tours
Université François Rabelais de Tours

Dr. Joseph C. Slater
Wright State University
Dayton, Ohio, USA

Dr. M. Sharaf
RWTH Aachen University
Aachen, Germany

Dr. Vijay K. Vasudevan
University of Cincinnati
Cincinnati, Ohio 45221, USA

OVERVIEW I

Fatigue of Materials II Advances and Emergences in Understanding
Edited by: T.S. Srivatsan, M. Ashraf Imam, and R. Srinivasan
TMS (The Minerals, Metals & Materials Society), 2013

FUNDAMENTALS OF FATIGUE CRACK INITIATION and PROPAGATION: SOME THOUGHTS

C. S. Pande[1]
Materials Science and Technology Division
Naval Research Laboratory
4555 Overlook Avenue S.W.
Washington DC 20375-5343, U. S. A.

Abstract

The prediction of fatigue properties of structural materials is rightly recognized as one of the most important engineering problem. Therefore a basic understanding of the fundamental nature of fatigue crack initiation and growth in metals has long been a major scientific challenge starting with the first dislocation model of fatigue crack growth of Bilby et al. in 1963. For this purpose understanding the process of emission of dislocations from cracks, and determining precise expressions for the size of the plastic zone size, the crack-tip opening displacement and the energy release rate of the cracks are some of the major technical challenges. In this short paper we comment briefly on some of our important recent results obtained theoretically and by in-situ TEM studies and discuss how they may contribute to the understanding of the phenomena

Keywords: fatigue, metallic materials, dislocations, cracks shielding.

Introduction

The progressive failure of a material by the incipient growth of flaws under cyclically varying stress is termed as fatigue and has over the years accounted for a vast majority of in-service failures in structures and components in the domain of aerospace engineering (airframe structures), in civil engineering (off-shore platforms, buildings, bridges), and in mechanical engineering. Such a failure can be a consequence of pure mechanical loading, aggressive environments (corrosion-fatigue) or elevated temperatures (creep-fatigue). The fatigue failure process can be categorized into the following discrete, yet mutually related and interactive, phenomena involving (i) cyclic plastic deformation prior to crack initiation, (ii) initiation of one or more microscopic cracks, (iii) growth and eventual coalescence of the microscopic cracks to form one or more macroscopic cracks, (iv) subsequent growth or propagation of both the microscopic and macroscopic cracks, and (v) final catastrophic failure [1].

Five fundamental questions need to answered in the study of the fatigue phenomenon are: (i) how cracks are produced? (ii) Where do the dislocations come from? (iii) How do the cracks interact with the dislocations, spanning both mobile and immobile? (iv) How cracks are able to propagate at loads far less than that needed for fracture? and (v) What is the effect and/or contribution of intrinsic microstructural features, such as: voids, interfaces, grain boundaries, and second-phase particles? Once these questions have been addressed, we will be in a position to use this knowledge to predict the fatigue behavior in a variety of materials and structures.

Crack Nucleation:

Crack initiation is the formation process of fine microscopic cracks having less than detectable length, say 0.001 inch or 0.025 mm. Over the years, research in the general area of metal fatigue and cyclic deformation has led to several hypotheses for fatigue crack initiation [2-15]. Most investigators generally agree that the fatigue cracks initiated at or near singularities on or just below the surfaces of metals. Such singularities may be inclusions, embrittled grain boundaries, sharp scratches, pits and slip bands [16]. However, subsurface nucleation has also been observed in metals having a strong adherent surface oxide, which retards crack initiation at the external surface [17-20]

It is now known that dislocations are convincingly involved both in the initiation and propagation of cracks in metallic materials. Basinski and Basinski [21] found that cracks could easily nucleate at the thin persistent slip bands (PSB's) that are generated immediately prior to fracture initiation. Mott [22] in 1958 suggested that vacancies are generated immediately below the surface and gradually accumulate, progressively grow under the influence of repeated loading and eventually coalesce to form fine microscopic cracks immediately below the surface. Antonopoulos and co-workers [23] extended the idea put forth by Mott [22] and proposed a model based on vacancy dipoles, which develop in the persistent slip band. Essmann and co-workers [24] in their independent study used similar ideas for the nucleation of fine microscopic cracks. However, it was Neumann [25] who developed and put forth a model based on an activation of two operating slip systems. Thus, it appears that the nucleation process of cracks is now fairly well understood for a wide spectrum of metals and we will not comment on this any further.

Dislocation Emission from Cracks

We have shown that once a crack has been nucleated or artificially produced, say by incorporating a notch, it acts as a source of dislocations. Several researchers have over the years in their experiments observed cracks that emitted dislocations at its tip. Ohr and coworkers [26-28] made an observation of the distribution of dislocations in the plastic zone during *in situ* tensile deformation of thin films obtained from the bulk in an electron microscope. A somewhat surprising result was that a careful observation of the region immediately ahead of the crack tip revealed this region to be free of dislocations. They called this region as Dislocation Free Zone (DFZ). Park and coworkers [29] and subsequently few others have made similar observations. Ding and coworkers [30] showed that by in-situ straining of pure tin solder foils in a transmission electron microscope the dislocations emanated from the blunted main crack tip and dislocation free zone were formed between the crack tip and the emitted dislocations. Pande (C. S. Pande, unpublished work) in his study obtained direct evidence of (i) dislocations being emitted at the crack tip, (ii) internal stresses due to a crack, and (iii) dislocation free zone, by straining a (111) oriented copper specimen (thinned from the bulk) using an in-situ transmission electron microscope.

Crack-Dislocation Interaction(s)

The next step is to be in a position to put to effective use this knowledge with the prime objective of predicting well-documented fatigue behavior. For an excellent summary of the fatigue phenomenon to including experimental observations, models and related theories reference is made to the material presented and discussed by Suresh [31]. Obviously, the first step towards both enabling and enriching our understanding of the fatigue phenomenon at the microscopic level is the need for an investigation of crack-dislocation interactions.

Most interesting work during the last few years has been development of discrete modeling approach by Pippan and coworkers [32-35], Deshpande and co-workers [36], and continuum modeling by Mastorakos and Zbib [37]. Hitherto, most of the documented studies related to modeling of the fatigue phenomenon have been two-dimensional in nature. However, Mastorakas and Zbib [37] used a three-dimensional analysis, which is more appropriate and realistic to facilitate a better understanding of the fatigue phenomenon. Several discrete dislocation simulations have been done in the past [32-52]. It has been found and documented that the threshold behavior can be related to the discrete nature of plastic deformation. Several fractographic features, to include abnormal striation spacing and zigzag propagation of the crack at low growth rates, can also be convincingly explained.

Masumura, Pande and Chou [53] have examined the case of two arrays of discrete edge dislocations in the presence of a semi-infinite crack for an elasto-static formulation for isotropic materials utilizing complex variable functions, they developed the forces that were required for equilibrium. Their analysis provided the following information: (i) number of dislocations to be in equilibrium with the crack at a given load, and (ii) the size of the dislocation free zone.

The resulting analysis also provided the following:

- (i) A measure of the size of the plastic zone.
- (ii) An estimate of the critical threshold for further dislocation emission.
- (iii) The magnitude of dislocation shielding of the crack.

These results accord well with the results of dislocation analysis reported by Lin and Thomson [54]. However, the method of Lin and Thomson cannot provide an expression for the size of the dislocation free zone because they replace the array of dislocations in front of the crack by two super dislocations.

In principle, the problem treated by Masumura and co-workers [53] can also be studied using the approximation of continuously distributed dislocations (CDDs). A continuum distribution of dislocations has been used in the past to study dislocation-crack interactions based upon the mathematical methods of singular integral equations. Bilby, Cottrell and Swinden (BCS) [55] modeled crack tip interaction using continuously

Distributed dislocations in lieu of a discrete number of dislocations. They were able to obtain an equilibrium distribution of dislocations both at and immediately around the crack tip (represented as a distribution of continuous dislocations). Their results were similar to that obtained and provided by Dugdale [56]. The analysis of BCS corresponded to the angle (θ) being 0 and size of the dislocation free zone being zero.

There are several problems that use the BCS model for studying the fatigue phenomenon and over the years various modifications have been suggested. A recent model using continuously distributed dislocations (CDDs) was put forth by Du and co-workers [57]. They investigated the dislocation free zone model for a symmetrical Mode I crack. Both the crack and two symmetrical plastic zones (inclined at angle θ the crack plane) by continuously distributed dislocations whose equilibrium position is considered by using a set of singular integral equations. For $\theta = 0$ the model easily reduces to that put forth by Chang and Ohr [58]. Although Chang and Ohr obtained closed form expressions for the various parameters, the model cannot be safely used for studying Mode I since they took the angle to be equal to zero. In contrast, Du and co-workers [57] consider the case of $\theta > 0$ but were not able to obtain any closed form solutions and were forced to resort to numerical computations for obtaining meaningful results.

The results of Masumura and co-workers [53] should be compared with that of Lin and Thomson [54]. The concept of a super dislocation was used by them to obtain a measure of K_D, provided a realistic measure of the mean position (r_m) could be easily determined. The super-dislocation is often used for classical dislocation pileups, but may not be a good approximation for an inverted pileup of dislocations. Some of the results obtained by Masumura and co-workers [53] are quite similar to those obtained by Lin and Thomson [54], thus providing a convincing verification of their test results. However, it should be noted that Masumura and co-workers [53] provide other information that is not possible in their treatment and this refers to the size of the plastic zone as a function of various loading conditions. They found a certain minimum value of k^* to be necessary for the emission of dislocations and its concurrent movement away from the crack. It is as yet not clear if this value of k^* can be associated with one of the two fatigue thresholds [59]. For further details the interested reader is referred to the paper by C. S. Pande that is contained in this volume.

Pippan and Weinhandl [60] pointed out that the simulation studies may open up many unanswered questions. These include the following:

(a) The effects of environment.
(b) A more detailed analysis of the three-dimensional nature of the crack.

(c) The role of internal dislocation sources and the crack tip as a source of dislocations.
(d) A transition from the blunting/re-sharpening crack propagation to a cleavage dominated process.

Crack Propagation

Many factors influence the rate of growth or propagation of fatigue cracks in metals, their alloy counterparts and composites based on metal matrices. During fatigue the driving force for crack growth is less than the driving force required for the same crack to grow under conditions of monotonic loading. A variety of theories have been proposed to rationalize fatigue crack behavior in the Paris law regime. In some cases, e.g. [61, 62], it is presumed that the fatigue crack growth rate is proportional to the cyclic crack opening displacement (COD) that implies a Paris exponent of two. The damage accumulation models give rise to a Paris exponent of four [63, 64]. More recent continuum plasticity-based models have been developed as detailed by Nguyen and coworkers [65] and Tvergaard, and Hutchinson [66], which can lead to a wider range of behavior.

Two structural factors that tend to affect or influence the glide processes taking place during crack growth are: (i) crystallographic orientation, and (ii) stacking fault energy. The effects of crystallographic orientation have been studied for single crystals of aluminum, and thus related to stacking fault energy with specific reference to polycrystalline copper [67]. Since the early mechanism proposed by Mott [22] is believed to contain the basic elements involved in crack initiation, a modification was made to include the primary growth stage [67]. The mechanism of growth in the second stage was strongly influenced by the normal stresses and the presence of substructure [68, 69]. Explanations dealing with the growth of crack along the sub-boundaries have associated this tendency with preferred crackling within the sub-boundaries themselves. Examination of thin films of aluminum in an electron microscope after cyclic loading [70] have shown that the boundaries are not regular arrays of dislocations as in the case of a simple tilt or simple twist boundary, but instead they consisted of complex tangles of dislocations.
We now consider some more recent studies pertaining to the crack propagation mechanisms. Jono and co-workers [71] observed growth behavior of the fatigue crack and slip deformation both at and near the crack tip by using an atomic force microscope. For a grain-orientated 3% silicon iron under conditions of constant amplitude loading, they found that in the lower ΔK region there was only one preferential slip system of this material in operation and the fatigue crack tended to grow along the slip plane. Constraints in slip deformation due to cyclic strain hardening resulted in either crack arrest or crack branching. However, in the high ΔK region two preferential slip systems operated simultaneously to an almost identical extent and the fatigue crack tended to grow in a direction perpendicular to the far-field load axis. The slip distance in one complete load cycle was measured quantitatively using the image processing technique. This observation is significant and points to an acceptable mode of fatigue crack propagation. Ihara and Tanaka [72] developed a mechanism for Mode I fatigue-crack propagation, using an idea which involved the initiation and opening of the cleavage-

mode crack. Their approach used a stochastic damage-accumulation model for gradual propagation of the fatigue crack. The calculated results for da/dN (crack growth rate per cycle) appeared to agree with the experimental data.

Once a microcrack or a void is produced in front of the crack it can be shown mathematically that the stress intensity needed for crack propagation is noticeably reduced. The reduction in stress intensity is substantial when the microscopic crack or void is close to the original crack (unpublished work by N. Louat and C. S. Pande).
In our opinion a detailed mechanism of crack propagation is one of the major challenge in the study of fatigue. Below we provide some new speculative ideas on this topic that is currently being in the process of development by others and us.

It is now well established that a maximum applied stress intensity factor, K_{max}, must exceed a certain value K_{th}, before crack propagation is possible. Another aspect of the problem is the reduction in fatigue crack growth and its eventual stoppage if DeltaK (the difference between K_{max} and K_{min}) falls below a threshold value. Thus, as clearly documented by Vasudevan *et al.* [73] in their recent review of a large body of fatigue data, there are two separate constraints or threshold conditions on fatigue growth. At high K_{max}, must be above some critical value and at low K_{max}, K_{max} must exceed some critical value K_{th} for a crack to propagate at a given rate. Both conditions must be satisfied at all times.

Experimental results determining these thresholds for various parameters such as microstructure, load ratio R ($R = K_{min} / K_{max}$), load history, overloading and environmental is now available. Hence any model of fatigue must address these issues in a quantitative fashion.

Our aim is to provide a quantitative description of these thresholds and to relate them to various parameters mentioned above. The model for fatigue that we want to propose is shown schematically below:

The micro crack of length *c* is potentially moving to the right under the action of the applied stresses. We first show that the crack cannot move until K_{max} exceeds a certain value. We also assume that during the fatigue process, a small micro crack of size *a* is created in front of the macro crack at a distance *delta* away. Both *a* and *delta* are expected to be small. When the cracks move they eventually join together, thus sharpening the macro crack tip for further advance by the repetition of this process. The validity of this model needs to be considered in detail and should be checked experimentally.

We first calculate, using this model the minimum threshold on K_{max} for the macro crack to move in absence of the micro crack and then calculate the effect of the minicrack on this value.
For simplicity, we initially consider only Mode I loading and the macro crack for mathematical purposes will be considered to be semi-infinite. Once the crack moves, the micro crack joins the macro crack, sharpening it in the process.

Under dynamic conditions (*i.e.,* crack moving), it can be shown that a relation exists between the fracture energy, *T,* and the stress intensity factor, *K*. The relation is

$$T = \frac{1}{E} B K^2, \tag{1}$$

where *B* is a function of crack velocity *v* only and *E* is Young's modulus and *K* is related to the crack length *c* and the stress field, by the usual relation,

$$\sigma_{ij} \approx K\, c\, \frac{K_o}{\sqrt{2\pi r}}, \tag{2}$$

where K_o is a function of angle crack is moving, all of these parameters may be a function of velocity. For small velocities, it can be shown that the function of *K* and *B* can be resolved as a product of a velocity term and a static term ($v = 0$). Thus,

$$K = k(v) K^s(c) \tag{3}$$

and

$$B = (1 - \frac{v}{v_o}). \tag{4}$$

where v_o is the maximum velocity (= acoustic) and superscript "s" denotes the static term ($v = 0$), *i.e*., independent of velocity. We also assume that *T* is independent of velocity.

Using equations (1), (3) and (4),

$$v = \frac{dc}{dt} \approx v_o \left[1 - \frac{E\,T}{\left(K^s\right)^2} \right] \tag{5}$$

where $v = (dc/dt)$ is crack velocity. Equation (5) shows that

$$v = 0 \text{ if } \left(K^s\right)^2 = E\,T \tag{6}$$

i.e., no crack motion if $K^s < (E\,T)^{1/2}$. In this calculation, K^s threshold is entirely dependent on the material parameters *E* and *T* and not much affected by microstructure, *etc*. This will be modified somewhat in the presence of the micro crack.

Calculations show that in the presence of a micro crack K^s threshold will increase from the value given by equation (6). A detailed calculation shows that the K^s value can more than double, making it easier for the crack to propagate.

The presence of the micro crack is then responsible for crack to propagate in fatigue, which it would not do otherwise. The threshold value in the presence of the micro crack is given by

$$K_F^g \approx K^g \left[\frac{\Delta + a}{\Delta} \frac{E_o(1 - \frac{\Delta}{\Delta + a})}{K_o(1 - \frac{\Delta}{\Delta + a})} \right] \tag{7}$$

where E_o and K_o are complete elliptic integrals of the first and second, respectively. Similarly that a second threshold is necessary. This one is required to initiate a micro crack that is used in the calculation discussed above. Needless to say these ideas need to be developed further.

Both the applied K field and the intrinsic friction stress affect the dislocation shielding. The interaction between these two parameters is inhomogeneous and non-linear and such scaling becomes an important issue. The concept of a super dislocation is useful to obtain a measure of K_D, provided that a realistic measure of the mean position ρ_m can be determined. The use of an inverted pile-up may not be a good approximation to these arrays where crack-dislocation interaction plays an important role.
We find that a certain minimum value of K^* is necessary for dislocation emission and its movement away from the crack. It is not yet clear if this value of K^* can be associated with one of the two fatigue thresholds.

Role of Surfaces

Recently, Pande Masumura and Chou (unpublished) considered the role of boundary displacements of a dislocation in the process of propagation of fatigue cracks. It was noticed during that investigations that information on boundary displacements produced by defects such as dislocations seems to be extremely sparse. They present (1) some supplementary remarks on the basis of the link between the harmonic functions and the method of images, and (2) a set of explicit expressions for the displacement field of an edge dislocation interaction with a boundary surface. They proceed by presenting a simple proof for the link between the harmonic functions and image field. The proof given by them seems simple, almost trivial, but it is new. It was not needed by early workers of electrostatics since their main interest was in spherical harmonics. But for defect fields in general, one needs a quantitative basis for the link in order to obtain stress functions which are not always harmonic. Surfaces and interfaces modify the stress fields of dislocations and their contributions should be carefully taken into account.

Concluding Remarks

Recently, the development and emergence of several analytical techniques has definitely stimulated the possibility of arriving at a basic understanding of crack initiation processes. It is truly hoped that further research along these lines coupled with theory and simulations will soon lead to a much better understanding of the complex crack initiation phenomenon and subsequent growth or propagation of the crack through the microstructure.

ACKNOWLEDGEMENTS

This work is based partly on an earlier review by the author and his coworkers. (C.S. Pande Tirumalai Srivatsan, Ashraf Imam,, Crack- dislocation interaction and its implication for a fundamental understanding of Fatigue Proceedings: International Symposium on Fatigue of Materials Advances and Emergences in Understanding Editor(s): Tirumalai Srivatsan, Ashraf Imam) and is supported by the Office of Naval Research. Finally are grateful to Dr. Robert A. Masumura for his substantial contributions to this work.

References

1. M. E. Fine: *Metallurgical Transactions* 11A, 1980, 365.
2. N. Thompson, and N. J. Wadsworth: Metal Fatigue Advanced Metal Physics (Philosophical Magazine Supplement 7, 1958, 72-82.
3. D. H. Avery and W. A. Backofen: Fracture of Solids (edited: D. C. Drucker and J. J. Gilman), Interscience Publishers, New York, USA, 1963, p. 339.
4. J. C. Grosskreutz and P. Waldow: *Acta Metallurgica*, Vol. 11, 1963, p. 717-727.
5. K. U. Snowden: *Acta Metallurgica*, Vol. 11, 1963, pp. 675-685.
6. C. Laird and G. C. Smith: *Philosophical Magazine*, Vol. 8, 1963, 1945-1955.
7. J. C. Grosskreutz: Sagamore Materials Conference: Fatigue: An Interdisciplinary Approach edited by J.J. Byre and N. L. Reed and V. Weiss, Syracuse University Press, 1964, p. 27-35.
8. J. C. Grosskreutz and G. G. Shaw: Fracture 1969 Proceedings of the Second International Conference on Fracture, Brighton, 12969, Chapman and Hall Publishers, London, UK, 1969, pp. 602-622.
9. W. J. Plumbridge and D. A. Ryder: *Metallurgical Reviews*, 136, 1969, pp. 119-139.
10. J. C. Grosskreutz: Phys. Status Solidi, 1970.
11. D. J. Duquette: Corrosion Fatigue: Chemistry, Mechanics and Microstructure, NACE 2, edited by O. F. Devereux, 1972, pp. 12-22.
12. S. Suresh and R. O. Ritchie: *International Metals Reviews*, 29 (6), 1984.
13. H. Mughrabi, R. Wang, K. Differt and U. Essmar: Fatigue Crack Initiation hy Cyclic Slip Irreversibility in High Cycle Fatigue," in Quantitative Measurement of Fatigue Damage, Dearborn, Michigan, 1982.
14. T. S. Srivatsan and T. S. Sudarshan: *Journal of Materials Science*, Vol. 23, 1988, 1521-1533.

15 A. S. Basinski, R. Pascual and S. J. Basinski: *Acta Metallurgica*, 31, 1983, 591-601.

16. M. E. Fine and R. O. Ritchie: Fatigue and Microstructure (edited by M. Meshii) American Society for Metals, Metals Park, Ohio, USDA, 1979, pp. 245-255.
17. P. H. Frith: *Journal of Iron and Steel Institute*, 159, 1948, 385-390.
18. W. A. Wood: Fatigue in Aircraft Structures, Academic Press, New York, USA, 1956.
19. H. E. Frankel, J. A. Bennett and W. A. Pennington: *Transactions of ASM*, 52, 1960, pp. 643-663.
20. W. A. Wood, S. Coustand and K. R. Sargart: *Acta Metallurgica*, 11, 1963, pp. 643-653.
21. Z. S. Basinski and S. J. Basinski. Low amplitude fatigue of copper single crystals-III. PSB sections. *Acta Metallurgica*, 33:1319–1327, 1985.
22. N. F. Mott. A theory of the origin of fatigue cracks. *Acta Metallurgica*, 6:195–197, 1958.
23. J. G. Antonopoulos, L. M. Brown and A. T. Winter, *Philosophical Magazine* 34, 549-563, 1976.

24. K. Differt, U. Essmann, and H. Mughrabi. A model of extrusions and intrusions in fatigued metals - II. surface roughening by random irreversible slip. *Philosophical Magazine A*, 54:237–258, 1986.
25. P. Neumann. Coarse slip model of fatigue. *Acta Metallurgica*, 17:1219–1225, 1969.
26. S. Kobayashi and S. M. Ohr, 37th Ann. Proc. Electron Microscopy Soc. of America", G.W. Bailey, Ed., San Antonio, TX, 1979, pp424-425.
27. S. Kobayashi and S. M. Ohr, "In-situ Fracture Experiments in B.C.C. Metals", *Philosophical Magazine,* 1980:A42: 763-772.
28. S. Kobayashi and S. M. Ohr, *Scripta Met*. "In-situ Observations of the Formation of Plastic Zone Ahead of a Crack Tip in Copper", *Scripta Metallurgica*, 1981:15:343-348.
29. C. G. Park C. S. Lee and Y. W.Chang Mechanical behaviour of materials Volume4, Proceedings of the 6th International Conference, Kyoto, Japan, July 29-Aug. 2, 1991. Vol. 4 (A93-40776 16-39), p. 3-9.
30. Ying Ding, Chunqing Wang, Mingyu Li and Weiqiang Wang, *Materials Science and Engineering B* 127 (2006) 62–69.
31. S. Suresh, "Fatigue of Materials", 2nd Edition, Cambridge University Press, Cambridge.
32. R. Pippan, *Acta Metall Mater* 39 (1991), pp. 255–262.
33. R. Pippan, *International Journal of Fracture,* 58 (1992), pp. 305–318.
34. F. O. Riemelmoser, R. Pippan and H. P. Stüwe, *International Journal of Fracture,* 85 (1997), pp. 157–168.
35. F. O. Riemelmoser, R. Pippan and H. P. Stüwe, *Acta Materialia* 46 (1998), pp. 1793–1799.
36. W. A. Curtin, V. S. Deshpande, A. Needleman, E. Van der Giessen, M. Wallin, International Journal of Fatigue 32 (2010) 1511–1520.
37. I. N. Mastorakos and H. M. Zbib, *Journal of Metals*, April 2008, p 59.
38. A. J. Wilkinson and S. G. Roberts, *Scripta Mater* 35 (1996), pp. 1365–1371.
39. A. J. Wilkinson, S. G. Roberts and P. B. Hirsch, *Acta Metall Mater* 46 (1998), pp. 379–390.
40. V. Doquet, *Fracture Engineering Materials Structures* 21 (1998), pp. 661–672.
41. E. van der Giessen, V. S. Deshpande, R. P. Cleveringa and A. Needleman, *J Mechanics Physics of Solids* 49 (2001), pp. 2133–2153.
42. V. S. Deshpande, A. Needleman, E. van der Giessen, *Acta Materialia* 49 (2001), pp. 3189–3203.
43. R. Pippan, R. O. Riemelmoser, H. Weinhandl and H. Kreuzer, *Philosophical Magazine* 82 (2002), pp. 3299–3309.
44. R. Pippan and R. O. Riemelmoser, *Z Metallkunde* 86 (1995), pp. 823–826.
45. F. O. Riemelmoser, P. Gumbsch and R. Pippan, *Material Transactions* 42 (2001), p. 2.
46. F. O. Riemelmoser, R. Pippan and H. P. Stüwe, *Acta Materialia* 46 (1998), p. 1793.
47. P. Hansson and S. Melin, *Engineering Fracture Mechanics* 75 (2008), pp. 1400–1411.
48. C. Bjerkén and S. Melin, *International Journal Fatigue*, 25 (2003), pp. 559–566.

49. C. Bjerkén and S. Melin, *Engineering Fracture Mechanics* 71 (2004), pp. 2215–2227.
50. P. Hansson and S. Melin, *International Journal of Fatigue* 27 (2005), pp. 347–356.
51. S. Groh, S. Olarnrithinum, W. A. Curtin, A. Needleman, V. S. Deshpande and E. van der Giessen, *Philosophical Magazine*, 88 (2008), pp. 3565–3583.
52. B. Kunkler, O. Duber, P. Koster, U. Krupp, C. P. Fritzen and H. J. Christ, *Engineering Fracture Mechanics*, 75 (2008), pp. 715–725.
53. R. A. Masumura, C. S. Pande and Y. T. Chou, International Journal of Fatigue, Volume 27, Issues 10-12, October-December 2005, Pages 1170-1174.
54. I. H. Lin and R. Thomson, "Cleavage, Dislocation Emission, and Shielding for Cracks under General Loading", *Acta Metallurgica,* 1986:34:187-206.
55. B. A. Bilby, A. H. Cottrell, and K. H. Swinden, "The Spread of Plastic Yield From a Notch", *Proc. R. Soc. London. Ser.*, 1963:A272:304.
56. D. S. Dugdale, *Journal Mechanics Physics of Solids*, 1960:8:100.
57. K. -L. Du, K. L, J. -B. Lu, and R. -W. Li, "A Continuous Dislocation Model of Mode I Crack", 15th ASCE Engineering Mechanics Conference, Columbia University, NY, 2002, pp.1-8.
58. S. J. Chang and S. M. Ohr, "Dislocation-free Zone Model of Fracture", *Journal Applied. Physics*, 1981:52:7174-7181.
59. A. K. Vasudevan, K. Sadananda and G. Glinka, "Critical Parameters for Fatigue Damage", *International Journal of Fatigue*, 2001:23:S39-S53.
60. R. Pippon and H. Weinhandl, International Journal of Fatigue 32 (2010) 1503–1510.
61. C. Laird, G. C. Smith, Crack propagation in high stress fatigue. Philos Mag, 1962; 7:847–57.
62. F. A. McClintock: Plasticity aspects of fracture. In: H. Liebowitz, editor. Fracture: an advanced treatise, vol. 3. New York: Academic Press; 1971. p. 47–225.
63. J. Weertman:. "Rate of growth of fatigue cracks calculated from the theory of infinitesimal dislocations distributed on a crack plane. In*ternational Journal of Fracture*, 1966; 2:460–7.
64. J. R. Rice Mechanics of crack tip deformation and extension by fatigue. Fatigue Crack Propagation, ASTM STP 1974; 415: 247–309.
65. O. Nguyen, A. Repetto, M. Ortiz, A. Radovitzky, A cohesive model of fatigue crack growth. *International Journal of Facture* 2001; 110: 351–69.
66. V. Tvergaard, J.W. Hutchinson. Crack growth per cycle by blunting and void growth. In: A. F. Blom, editor. Proceedings of the 8th international Conference on Fracture. vol. 1. West Midlands (UK): EMAS; 2002. p. 107–16.
67. A. J. McEvily and R. C. Boettner: *Acta Metallurgica*, Vol. 11, 1963, 725-743.
68. J. C. Grosskreutz: *Journal of Applied Physics*, 33, 1962, 1787.
69. J. Holden: *Philosophical Magazine*, 6, 1961, 547-557.
70. J. C. Grisskreutz: Proceedings of the Fifth International Congress for Electron Microscopy, Academic Press, New York, 1962, J-9.

71. M. Jono, A. Sugata and Y. Uematsu, Atomic force microscopy and the mechanism of fatigue crack growth, *Fatigue Fracture Engineering Materials Structures,* 24 (2001), pp. 831–842.
72. C. Ihara and T. Tanaka, *Engineering Fracture Mechanics*, Volume 74, 2007, pp 1488-1498.
73 A.K. Vasudevan, K. Sadananda and G.Glinka, "Critical Parameters for Fatigue Damage", *Int. J. of Fatigue*, 2001:23:S39-S53.

Fatigue of Materials II Advances and Emergences in Understanding
Edited by: T.S. Srivatsan, M. Ashraf Imam, and R. Srinivasan
TMS (The Minerals, Metals & Materials Society), 2013

IN SITU THREE DIMENTIONAL (3D) X-RAY SYNCHROTRON TOMOGRAPHY OF CORROSION FATIGUE IN Al7075 ALLOY

Sudhanshu S. Singh[1], Jason J. Williams[1], X. Xiao[2], F. De Carlo[2] and N. Chawla[1]

[1] Materials Science and Engineering, Arizona State University, Tempe, AZ 85287-6106, USA

[2] Advanced Photon Source, Argonne National Laboratory, Argonne, IL, USA

Keywords: Al7075, X-ray tomography, EXCO, fatigue crack growth, bubbles, corrosion products

Abstract:

X-ray synchrotron tomography was used to investigate the fatigue corrosion behavior of Al7075-T651 alloys in a corrosive solution. Single edge- notch specimens machined along the L-T orientation were used for corrosion-fatigue testing, under a load ratio of R = 0.1. The evolution of reaction between the corrosive fluid and the metal was observed, as well as the formation of hydrogen bubbles. The shape of these bubble also changed significantly during the fatigue cycle. The fatigue crack growth rate was found to be significantly higher in the corrosive solution, compared to crack growth in air, for stress intensity levels between 5 and 10 MPa$\sqrt{}$m.

Introduction

7075 aluminum alloys are used extensively due of their high strength to weight ratio [1-3]. However, these alloys are susceptible to corrosion and show a significant decrease in fatigue life when exposed to a corrosive environment [4]. In the presence of a corrosive fluid, fatigue crack growth rate increases significantly compared to crack growth rates in air [5, 6]. X-ray tomography is an excellent technique for studying the deformation behavior of materials in three dimensions (3D). It is non-destructive, which enables a time dependent (4D) understanding of the material behavior [7]. X-ray tomography has been used to understand fatigue crack growth behavior in Al-Mg-Si [8, 9], Al7075 [7, 10] and Mg alloy [11] in air. The *in situ* corrosion fatigue behavior of 7075 aluminum alloys has not been studied by x-ray tomography.

In this study, we have used X-ray synchrotron tomography to study the fundamentals of corrosion-fatigue in Exfoliation Corrosion (EXCO) solution. The reaction between corrosive fluid and the aluminum alloy, shape of bubbles, and crack growth rate were studied.

Materials and Experimental Procedure

The material used in this study was a commercially available rolled 7075-T651 aluminum alloy (Al–6Zn–3Mg–2Cu). Details of the microstructure of this alloy can be found elsewhere [12]. As shown in Figure 1, single edge-notch specimens were machined (25 mm x 2.8 mm x 0.8 mm) by electro discharge machining (EDM) such that the rolling axis (L-T orientation) was parallel to the loading axis. Fatigue pre-cracking was performed *ex situ* on a microforce testing system (MTS Tytron 250) in tension-tension fatigue (4 Hz, R=0.1, $a_0 \approx 1$ mm, $\Delta K \approx 5$ MPa$\sqrt{m}$).

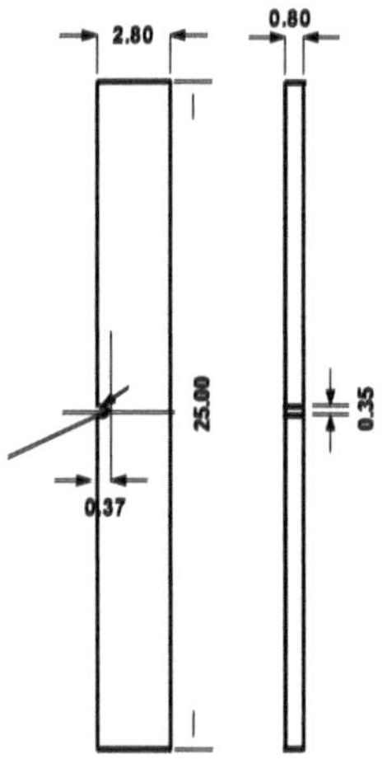

Figure 1: Single edge notch specimen used for *in situ* experiments (dimensions are in mm).

The loading stage for *in situ* corrosion-fatigue experiments using X-ray synchrotron tomography is shown in Figure 2(a). The load cell has a 500 N capacity, and the stepper motor has a Captive Linear Actuator capable of 8 μm displacement per step and 12 mm total stroke displacement. Fatigue experiments were conducted in the Paris law regime at a frequency of 0.5 Hz and R-ratio of 0.1. More details of loading stage can be found elsewhere [10]. Figure 2b shows the specimen in a bath with the EXCO solution (4M NaCl, 0.5 M KNO_3 and 0.1 M HNO_3). The specimen was clamped at both ends inside the loading stage (Fig. 2a) for *in situ* corrosion-fatigue experiments.

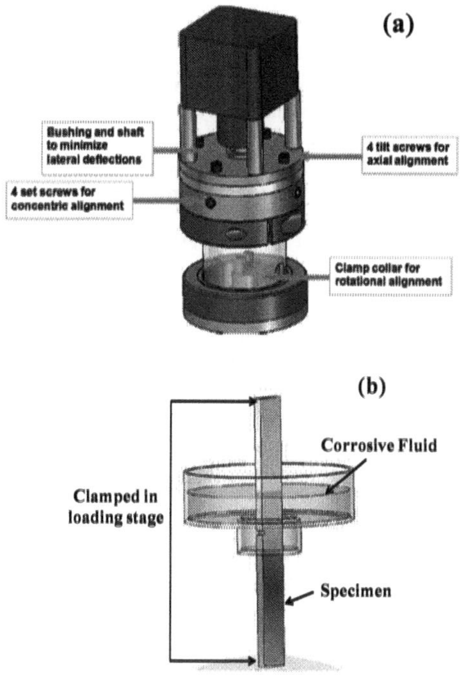

Figure 2: *In situ* corrosion-fatigue setup (a) loading stage and (b) corrosion system inside the loading stage.

X-ray microtomography measurements were carried out using the 2-BM beamline at the Advanced Photon Source (APS) at Argonne National Laboratory. Tomography was conducted *in situ* at various fatigue cycle increments, and at various points in the sinusoidal loading curve, for a given cycle. The x-rays penetrated a sample volume of approximately 1 mm x 2.7 mm x 1 mm. "Pink beam" was used for these experiments. While pink beam is polychromatic, the advantage is that it uses high photon flux (10^{14} Photons/s/mm^2) and low exposure time (100 μs), which considerably reduces the scanning time. This study used a scan time of 60 seconds with pink beam compared to 20 minute scans using a monochromatic beam in previous study [7]. Pink beam was focused on the specimen and a LuAG:Ce scintillator screen was used to convert the transmitted X-rays to visible light. This was coupled with an objective lens and a PCO Dimax CMOS camera to achieve pixel sizes of 2.2 and 1.46 μm (in two different sessions at APS). 2D projections were collected at angular increment of 0.12°. These 2D projections were then reconstructed in 3D using a filtered-back-projection algorithm. The grayscale images were segmented using thresholding using image analysis software (ImageJ, Bethesda, MD).

Results and Discussion

Figure 3 shows the comparison of fatigue crack growth rate (da/dN), versus stress intensity factor range (ΔK) of Al7075-T651 in EXCO solution and in air. The pixel size in these experiments was 2.2 μm. The data for fatigue crack growth in air are from a previous study [7]. It is evident that the crack growth rate is significantly higher in corrosive fluid than in air for stress intensity levels between 5 MPa√m and 10 MPa√m.

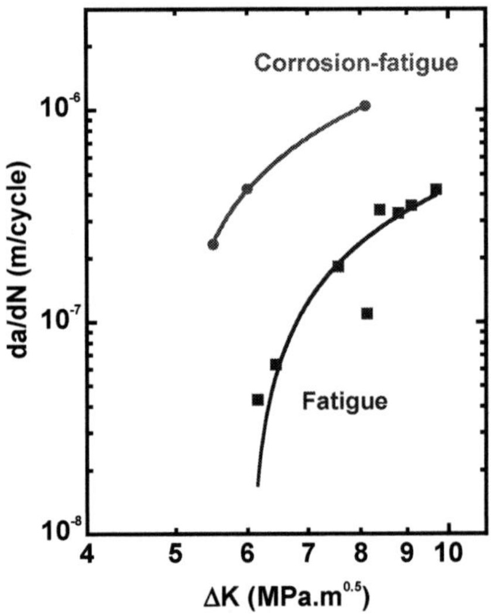

Figure 3: Comparison of fatigue crack growth rate in EXCO solution and in ambient air by *in situ* X-ray tomography. FCG rate increases significantly in EXCO solution.

X-ray tomography 2D images of damage due to exfoliation corrosion are shown in Figure 4a and 4b. Exfoliation corrosion is usually inter-granular corrosion along the elongated grain boundaries parallel to the rolling axis, as shown in Figure 4a. 7075 grain boundaries are more anodic because they are enriched in Mg and depleted in Cu compared to the grain interior which makes grain boundaries more susceptible to exfoliation corrosion [13]. These elongated grains are subsequently forced apart (grain lift out) by the volume expansion of the corrosion products (Fig.

4b), which allow corrosive fluid to penetrate deeper into the material. The corrosion product of aluminum is formed according to the following reaction [14]:

$$\text{Anode: } Al = Al^{3+} + 3e^-$$
$$\underline{\text{Cathode: } 3H^+ + 3e^- = 3/2\ H_2}$$
$$Al + 3H_2O = Al(OH)_3 + H_2\ (g)$$

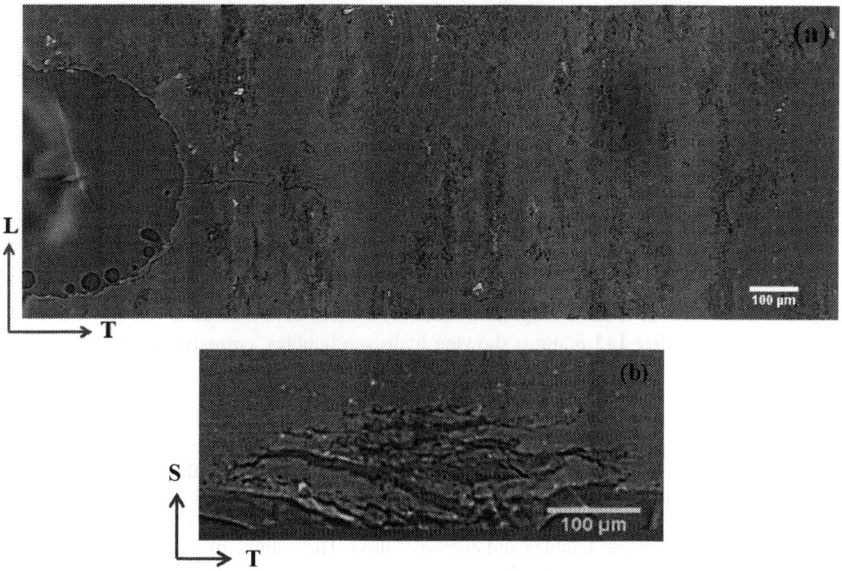

Figure 4: X-ray tomography 2D images showing (a) exfoliation corrosion parallel to rolling direction and (b) exfoliating as a function of depth from the surface.

Figure 5 show the side view of 2D slices with and without EXCO solution (142 minutes after addition of the solution). The pixel size in these experiments was 1.46 µm. In Figure 5b, bubbles and corrosive fluid inside the crack, and corrosion product surrounding the crack are visible which were not present before addition of fluid (Fig. 5a). The EDM notch is in the left side of the image. At this point, we hypothesize that these bubbles in Figure 5b are mostly hydrogen bubbles produced during the reaction between corrosive fluid and Al alloys.

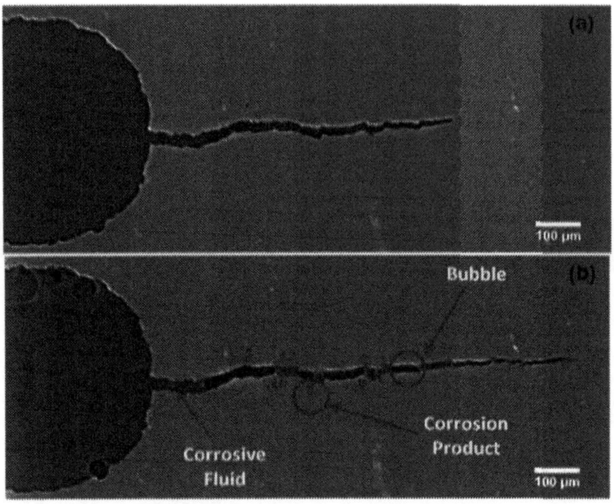

Figure 5: Side view of X-ray tomography slices (a) without EXCO solution and (b) with EXCO solution after 142 minutes showing hydrogen bubbles, corrosive fluid, and corrosion products.

A part of crack (150 out of about 490 2D slices) and bubbles inside it were segmented using thresholding in image analysis software (ImageJ, Bethesda, MD). Since the crack consists of bubbles and corrosive fluid, segmented bubbles were subtracted from the segmented crack to obtain corrosive fluid. The 3D reconstruction of a 2D stack of segmented images was performed using commercially available software (MIMICS, Ann Arbor, MI). Figure 6 shows the 3D reconstruction of the crack, bubbles and corrosive fluid. The volume of bubbles inside the crack was around 32% of the total crack volume.

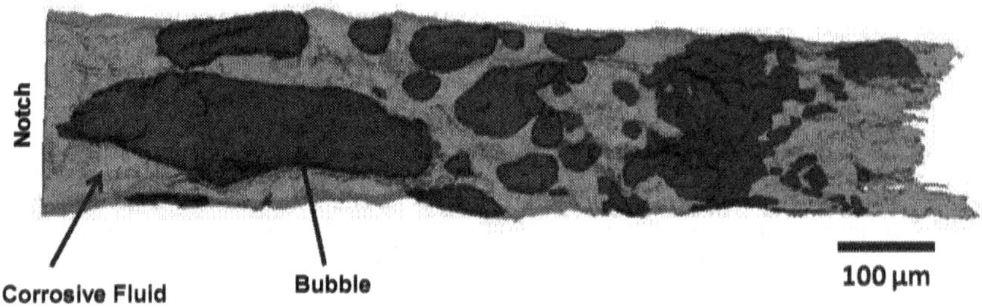

Figure 6: 3D reconstruction of crack (bubble + corrosive fluid) after 142 minutes in EXCO solution.

Corrosion products were also segmented using thresholding in ImageJ on a few slices (from the part of crack) as shown in figure 7. The corrosion products growing away from the crack are visible (also in Fig. 5b). The average height of the corrosion products from around 30 different measurements on 2D slices was found to be 35 μm.

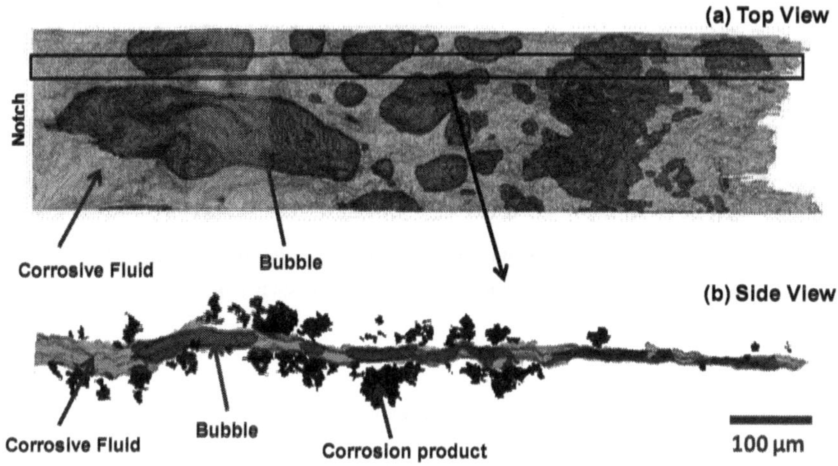

Figure 7: 3D reconstruction of (a) top view of the fatigue crack (bubble + fluid) and (b) corrosion products from selected area of the segmented crack after 142 minutes of addition of EXCO solution.

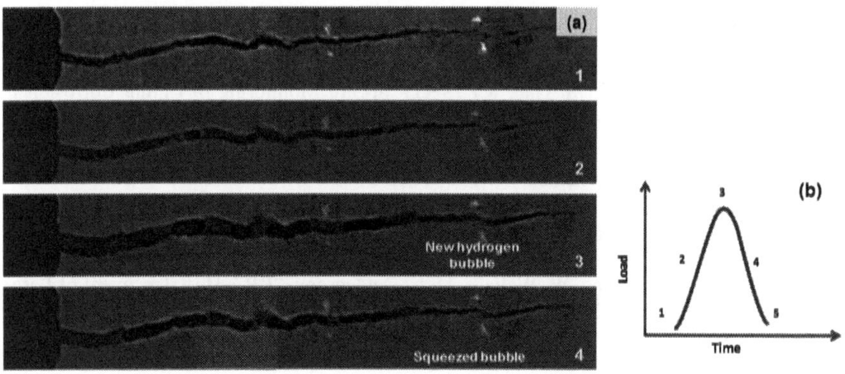

Figure 8: (a) 2D X-ray tomography images showing hydrogen bubble formation and change in morphology of bubbles in one fatigue cycle (b) corresponding fatigue cycle.

Figure 8 shows the changes in bubbles shape and formation of a new bubble during one fatigue cycle. A bubble formed at position 3 in a fatigue cycle (it is not present at position 2). All bubbles get squeezed as the crack closes during unloading at position 4.

Conclusions

In situ X-ray synchrotron tomography has been used to study corrosion-fatigue behavior in 7075-T651 alloy. The following conclusions can be made from this study:

(1) *In situ* X-ray tomography is an excellent technique to study three dimensional corrosion-fatigue behavior. The fatigue crack growth rate from 2D slices in corrosive fluid (EXCO solution) was compared with the growth rate in ambient air. The fatigue crack growth rate was found to be significantly higher in EXCO solution than in ambient air.

(2) *In situ* X-ray tomography also enables time dependent study (4D). Hydrogen bubbles form due to reaction between Al alloys and corrosive fluid and their morphology inside the crack changes with time.

Acknowledgements

The authors are grateful for financial support from the Naval Air Warfare Command through a subcontract through Technical Data and Analysis, under contract number N00421-10-P-0818 (N. Iyyer, TDA, and M. Kittur, NAVAIR, program managers). Use of the Advanced Photon Source was supported by the U.S. Department of Energy, Office of Science, Office of Basic Energy Sciences, under Contract No. DE-AC02-06CH11357.

References

[1] C. M. Liao, J. M. Olive, M. Gao, R.P. Wei, "In-Situ Monitoring of Pitting Corrosion in Aluminum Alloy 2024", *Corrosion*, 54 (6) (1998), 451-458.

[2] A. AlHazaa, T.I. Khan, I. Haq, "Transient liquid phase (TLP) bonding of Al7075 to Ti–6Al–4V alloy" *Materials Characterization*, 61 (2010), 312-317.

[3] C. M. Cepeda-Jimenez, J. M. Garcia-Infanta, O. A. Ruano, F. Carreno, "Mechanical properties at room temperature of an Al–Zn–Mg–Cu alloy processed by equal channel angular pressing" *Journal of Alloys and Compounds*, 509 (2011), 8649-8656.

[4] R. T. Holt, M. D. Raizenne, W. Wallace, D. L. DuQuesnay, "RRA Heat Treatment of Large Al 7075-T6 Components", National Research Council of Canada Ottawa (Ontario) Inst for Aerospace Research, (2000).

[5] P. S. Pao, M. A. Imam, L. A. Cooley, G. R. Yoder, "Comparison of Corrosion Fatigue of Al-Li Alloy AA2090-T8E41 and Alloy AA 7075-T651 in Salt Water", *Corrosion*, 45 (7) (1989), 530-535.

[6] T. B. Mills, "The combined effects of prior-corrosion and aggressive chemical environments on fatigue crack growth behavior in aluminum alloy 7075-T651", (Ph.D. thesis, The University of Utah, 1997).
[7] J. J. Williams, K.E. Yazzie, E. Padilla, N. Chawla, X. Xiao, F. De Carlo, "Understanding fatigue crack growth in aluminum alloys by in situ X-ray synchrotron tomography" *International Journal of Fatigue*, In Press.
[8] H. Zhang, H. Toda, P. C. Qu, Y. Sakaguchi, M. Kobayashi, K. Uesugi, Y. Suzuki, H. Zhang, "Three-dimensional fatigue crack growth behavior in an aluminum alloy investigated with in situ high-resolution synchrotron X-ray microtomography" *Acta Materialia*, 57 (2009), 3287-3300.
[9] H. Toda, S. Yamamoto, M. Kobayashi, K. Uesugi, H. Zhang, "Direct measurement procedure for three-dimensional local crack driving force using synchrotron X-ray microtomography" *Acta Materialia*, 56 (2008), 6027-6029.
[10] J. J. Williams, K. E. Yazzie, N. C. Phillips, N. Chawla, X. Xiao, F. De Carlo, N. Iyyer, M. Kittur, "On the correlation between fatigue striation spacing and crack growth rate: A three-dimensional (3-D) X-ray synchrotron tomography study" *Metallurgical and Materials Transactions A*, 42 (13) (2011), 3845-3848.
[11] A. King, W. Ludwig, M. Herbig, J.-Y. Buffiere, A. A. Khan, N. Stevens, T. J. Marrow, "Three-dimensional in situ observations of short fatigue crack growth in magnesium", Acta *Materialia*, 59 (2011) 6761-6771.
[12] A. Bonakdar, F. Wang, J. J. Williams and N. Chawla, "Environmental Effects on Fatigue Crack Growth in 7075 Aluminum Alloy", *Metallurgical and Materials Transactions A*, 43 (8) (2012), 2799-2809.
[13] A. Joshi, C. R. Shastry, M. Levy, "Effect of Heat-Treatment on Solute Concentration at Grain-Boundaries in 7075 Aluminum-Alloy", *Met Trans A*, 12 (6) (1981), 1081-1088.
[14] C. Vargel, Corrosion of Aluminum, Elsevier Inc, San Diego CA (2004).

Fatigue of Materials II Advances and Emergences in Understanding
Edited by: T.S. Srivatsan, M. Ashraf Imam, and R. Srinivasan
TMS (The Minerals, Metals & Materials Society), 2013

Variable Amplitude Fatigue

Narayanaswami Ranganathan[1] , Damien Joly and René Leroy[1]

[1] Laboratoire de Mécanique et Rhéologie ;
Université François Rabelais de Tours
Polytech Tours
7 avenue Marcel Dassault, 37200 Tours, France

Keywords : Fatigue crack growth, Variable amplitude tests, microstructure, fractography

Abstract

Fatigue crack growth behavior of selected aluminum alloys under variable amplitude loading is discussed in this study, based principally on experimental observations. The tests include single overloads tests in different environments, block load tests and tests using an aircraft wing loading spectrum. It is shown that conditions favoring a planar slip behavior lead to very high delays as opposed to conditions leading to multiple slip behavior. The Aluminium Liithium alloy studied here, has the best fatigue crack growth resistance in almost all test conditions studied here as compared to other conventional alloys. Under the spectrum loading studied here, the same alloy exhibits a change in micromechanism leading to a four fould acceleration of growth rates. Acceptable life predictions can be made, by taking into account this crack acceleration effect.

Introduction

Variable amplitude fatigue crack growth is a very important aspect, especially with respect to aircraft fatigue. The first pioneering studies were carried out in the Royal aircraft laboratories at Farnborough in the early 50s. One of the major accidents that led scientists to think of damage tolerance is the notorious Comet aircraft crashes. The development of Fracture Mechanics concepts and the introduction of Paris law and its derivatives gave a new impetus and gave rise to new ideas. Recent advances in the field of crack measurements with high resolution microscopy have led to suggestions that even the initial stages of small crack propagation can be analyzed in terms of fracture mechanics.

This paper addresses variable amplitude crack growth behavior under different loading conditions in aluminum alloys 2024 in the T351 condition, 7075 alloy in T351, T651 and T7351 conditions and the Aluminium Lithium alloy 8090 T651 .
Different kinds of variable amplitude considered are:
-Single overload tests
- Block load tests
- Tests under spectrum loading

The analysis of the results are done with an effort to highlight the intrinsic material behavior based on tests carried out in vacuum. The effect of material behavior as defined by precipitate coherency is discussed in terms of the results obtained in the 7075 alloy in the three heat treatment conditions.

Fractographic analyses are carried out to bring out salient aspects of micromechanisms, especially in the Aluminium Lithium alloy.

.

Materials studied and salient experimental details

Nominal composition and mechanical properties are given in tables 1a and 1b.

Table 1a - Composition (in %wt.) of the studied alloys (Al remaining)

Alloy	Zn	Mg	Cu	Cr	Fe	Si	Mn	Ti	Li
2024	0.04	1.5	4.46	0.01	0.22	0.1	0.66	0.02	-
7075	5.7	2.43	1.5	0.2	0.21	0.16	0.04	0.04	-
8090	-	0.55	1.14	-	0.05	0.03	0.1	0.03	2.34

Table 1b - Nominal mechanical properties

Alloy	E (GPa)	YS (MPa)	UTS (MPa)	A(%)
2024 T351	73	300	500	16
7075 T351	70	458	583	10.6
7075 T651	71.1	527	590	11
7075 T7351	70.9	470	539	11.7
8090 T651	81.2	430	480	13

All the studied materials show a pancake shaped grain structure: the average grain size (given in the ourder L,T and S directions) of the 2024 alloy is 190 x 60 x 70 µm. The grain size of the 7075 T351 alloy is 700 x 200 x 50 µm and that of the 7075 T7351 alloys is 600 x 190 x 40µm. Similar microstructure is observed for the other two alloys.

The 2024 alloy is annealed and aged at room temperature. The 7075 alloy is studied under three ageing conditions, RT ageing T351, peak-aged T651 and over-aged T7351 conditions. The 8090, Al Li alloy is studied in the peak aged condition.

The constant amplitude and variable amplitude tests described here were carried out using Compact Tension (CT) specimens, 75mm wide and 10mm or 12 thick

The constant amplitude and variable amplitude tests presented here were carried out in ambient air and in vacuum ($<10^{-3}$ Pa) and a N_2 environment containing traces water vapor. For overload tests, two baseline load ratios, R, of 0.1 and 0.5 were studied. 100% overload tests were carried out at a load ratio of 0.1 covering near threshold to mid ΔK values.The repeated overload tests were carried out at a baseline loading at R = 0.1 and every 999 cycles an overload of 70% is applied.

The transport aircraft wing spectrum studied consists of 1000 different flights and 22547 cycles respectively. Compressive loads were not applied and the minimum ground load level was set to about zero. The tests were conducted under computer control and the crack was monitored either optically or by crack gages with a precision on the order of 0.01 mm.

Experimental details can be found in [1-5].

Experimental results and analysis

Comparison between constant amplitude behavior and post overload delay

Figure 1 shows the constant amplitude behavior of the studied alloys in air. For the 7075 alloy only the T7351 condition is shown.

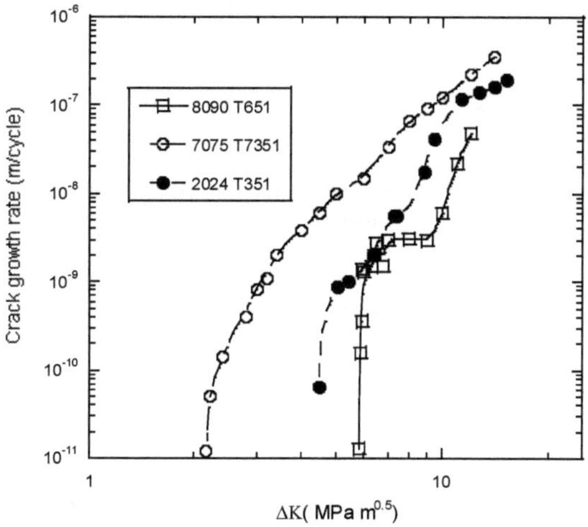

Figure 1 Constant amplitude behavior at r=0.1 in air

It can be seen here that the 8090 alloy has the best crack growth resistance throughout the ΔK range studied. The threshold values and measured crack growth rates at a ΔK of 10 MPa $m^{0.5}$, given in the following table illustrate the superior fatigue crack growth resistance of the Al Li alloy [4]

Table 2 Threshold values and growth rate at a ΔK of 10 MPa $m^{0.5}$

Alloy	8090 T651	2024 T351	7075 T7351
ΔK Kth (MPa $m^{0.5}$)	5.8	4.4	1.2
da/dN in m/cycle at ΔK = 10 MPa $m^{0.5}$	2 10^{-9}	1.8 10^{-8}	10^{-7}

In the following figure we compare the evolution of delay with respect to baseline ΔK for these three alloys.

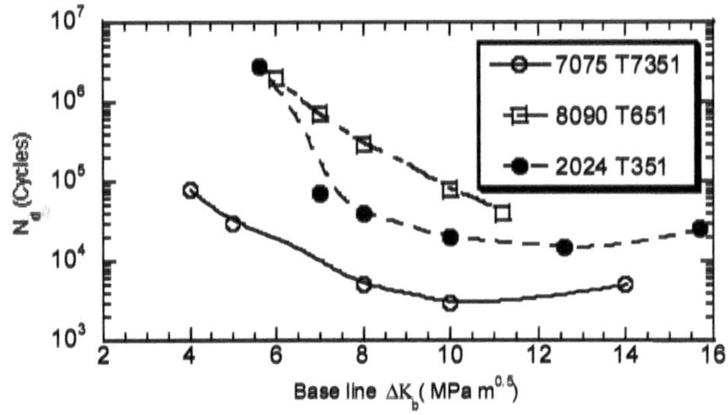

Figure 2 Number of delay cycles with respect to baseline ΔK – Tests carried out at a base line load ratio of 0.1

It can be seen here that, in the ΔK range studied, delay is systematically higher in the 8090 alloy as compared to the conventional alloys, except for the lowest ΔK studied. It should be mentioned that at the lowest ΔK_b , and only in the 2024 alloy, complete crack arrest was observed. The crack was considered to be arrested after 2 million cycles.
At a ΔK of 10 MPa $m^{0.5}$, delay is about 10^5 cycles in the Aluminium Lithium alloy; while it is only 3000 cycles in the 7075 T7351 alloy. In the 2024 alloy an intermediate delay of about 20000 cycles is observed.

It can be inferred from the above analysis that there is a good concordance between the constant amplitude fatigue crack growth resistance under constant amplitude loading and the post overload delay.

The effect material plastic behavior

The material plastic behavior in aluminum alloys depend upon precipitate coherency and the ambient environment [6]. This aspect is illustrated based on results on the 7075 alloys. In this alloy, the under aged T351condition is characterized by the presence of coherent precipitates. Precipitate coherency decreases with ageing conditions. In the peak aged T651 condition semi-coherent precipitates are found while in the over aged T7351 condition, precipitates are totally non coherent with the matrix. As a consequence, planar slip behavior is expected in the T351 condition and multiple slip in the over aged condition while in the T651 condition a mixed slip behavior is expected. Previous studies have shown that such effects are observed more readily in vacuum, as in air the presence of active molecules (OH or H) could enhance multiple slip activity.

The constant amplitude and post overload delay in the 7075 alloys studied in vacuum are shown in figures 3 and 4.

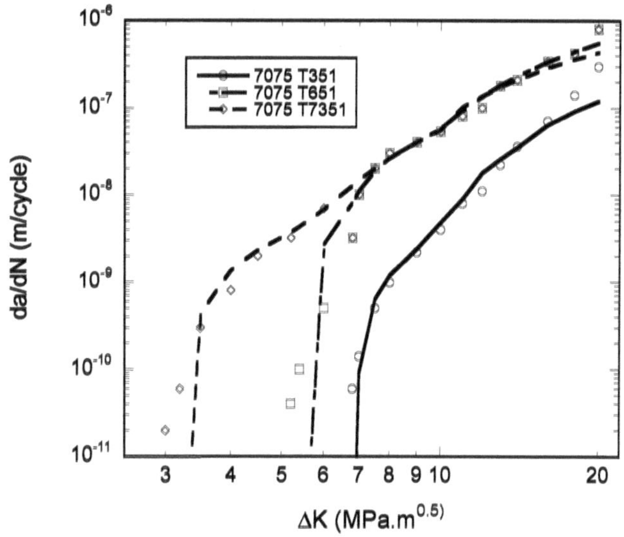

Figure 3 Constant amplitude behavior of the 7075 alloy in the three heat treatment conditions at a load ratio of 0.1 in vacuum

It can be seen that the threshold value is the highest for the T351 condition followed by T651 and the T7351 conditions.

In the studied ΔK range, the under aged condition has the best crack growth resistance . the behaviors of the T651 and T7351 conditions become similar for ΔK values higher than 8 $MPam^{0.5}$

It can be observed here that the delay in the T351 condition is the highest throughout the studied ΔK range and the T7351 condition has the lowest delay. In the T651 condition, high delays almost comparable with those in the under aged condition is observed at Low ΔK values, while at higher base line ΔK levels, lower delays are observed after a transition. It can again be seen that post overload delay is in agreement with constant amplitude behavior.

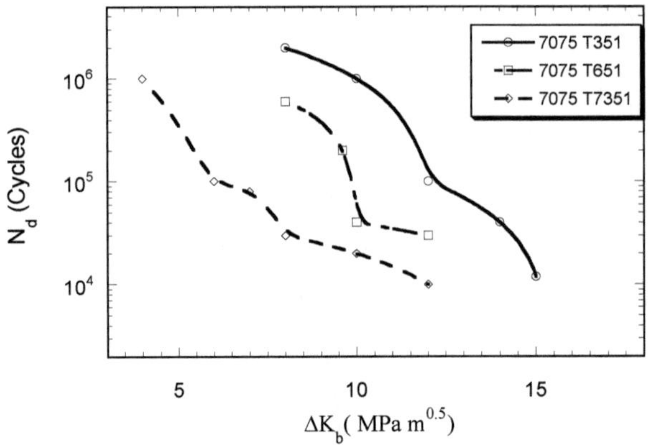

Figure 4 Delay behavior in vacuum for the 7075 alloy in three heat treatment conditions at a base line load ratio of 0.1

Repeated overload tests

These tests refer to block load tests consisting of 1000 base line load cycles at a load ratio of 0.1 followed by an overload of 170%. These tests average the accelerating effect of the overload itself with the deceleration effect after the overload. Results are presented in the next figure where the average growth rates are compared to the baseline ΔK for the 3 studied alloys .

It can be seen here that the 8090 alloy has the best crack growth resistance throughout the ΔK range studied. The differences with respect to the other two alloys are the highest for ΔK about 14 MPa $m^{0.5}$ with crack growth rates in the 7075 alloy being almost 10 times as much as that in the 8090 alloy. For the same ΔK, the crack growth rate in the 2024 alloy is about 6 times as that in the 8090 alloy. In fact the 8090 alloy exhibits a near constant growth rate in the range $10<\Delta K <15$ MPa $m^{0.5}$. At , high ΔK values the 8090 alloy has almost the same crack growth rate as the 2024 alloy.

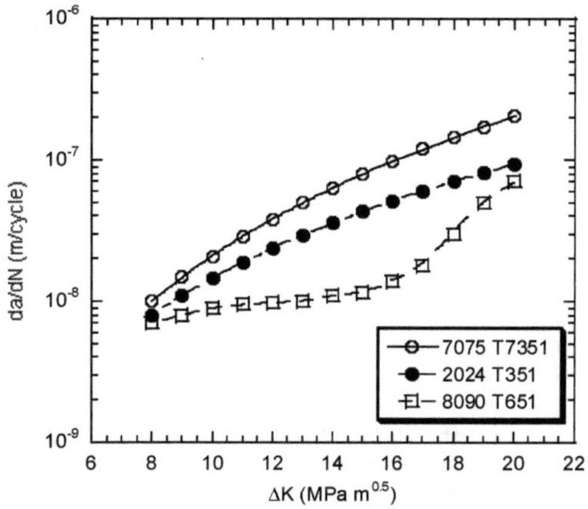

Figure 5 Crack growth behavior under repeated ol tests

Again we notice that there is a concordance between constant amplitude and variable amplitude behavior.

Crack growth and spectrum load tests

The variable amplitude spectrum studied here, was developed by the French aircraft industries in the 1980s and represents part of theoretical spectra used in experimental programs [] . It consist of 1000 different flights of different severity. Fig [6] shows a typical flight.

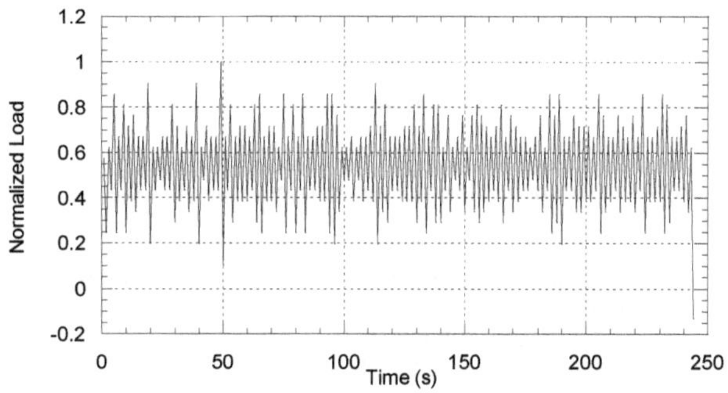

Figure 6 Typical flight

Crack length evolution measured under this spectrum are shown in the next figure. One can see here that under such conditions, the 2024 T351 alloy has the best crack growth resistance followed by the 8090 alloy and the 7075 alloy.

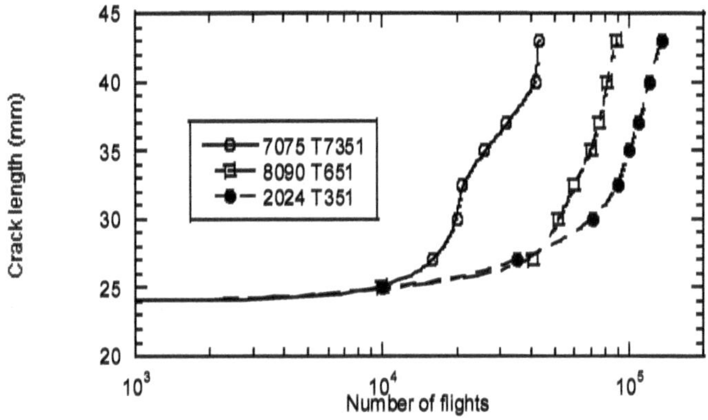

Figure 7 Crack length evolution versus number of flights for the three alloys

Table 3 Crack growth lives (crcak length range 24 to 42mm) measured for these tests

Alloy	8090 T651	2024 T351	7075 T7351
Life (flights)	89000	135000	43000

Discussion

It can be seen in this study, that there is a good concordance between constant and variable amplitude crack growth resistance in the studied aluminum alloys, except for tests under spectrum loading.

First of all let us discuss about crack growth mechanisms as inferred from fractography. In the 7075 or the 2024 alloy in the under aged T351 condition, planar slip is observed, typically for tests in vacuum. These leads to the formation of large crystallographic facets on the fracture surface, as shown the following photo, which shows an overload application point.

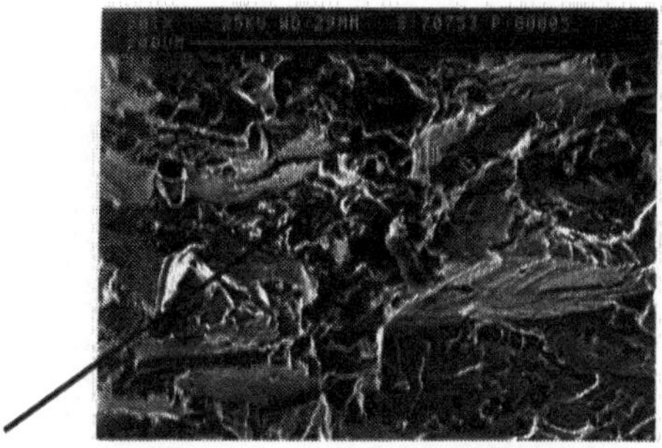

Figure 8a Fracture appearance after an overload in the 7075 T351 alloy in vaccum : Crack grows from left to right – The arrow shows the overload point

These facets can be identified to be along (111) planes [7,8].

In the case of multiple slip behavior, as fo rover aged alloys in vacuum and for most alloys tested in air, a flat fracture surface is observed as shown in the following figure.

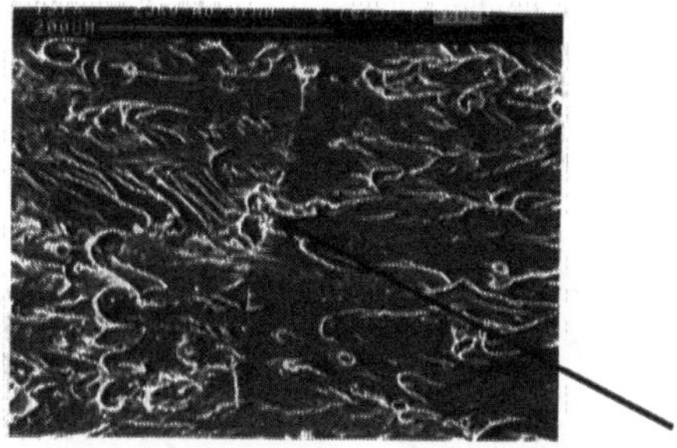

Figure 8b Fracture appearance after an overload in the 7075 T7351 alloy in air : Crack grows from right to left – The arrow shows the overload point

In the case of the Aluminium Lithium alloy , crystallographic facets are observed for tests in air as well and it is concomitant with high crack growth resistance under constant amplitude and post over load behavior. The following figure shows a chaotic fracture surface appearance after an overload in the 8090 alloy.

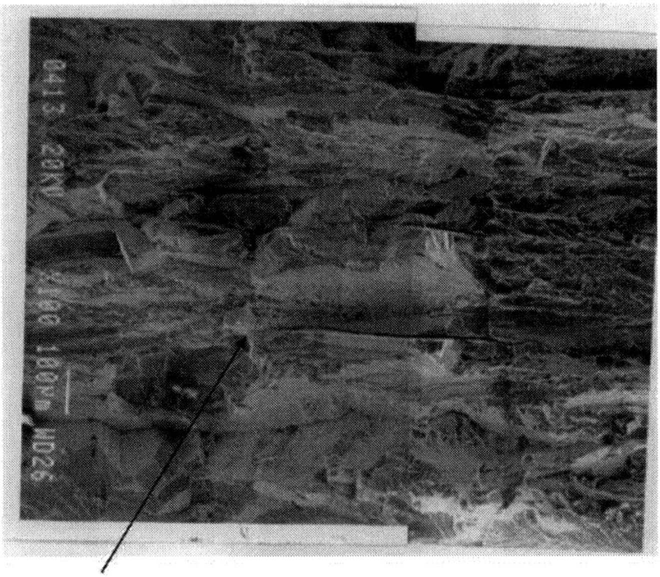

Figure 9 Overload appearance in the 8090 alloy in air. The crack growing from left to right and the arrow shows the overload point

Within these facets very small dimples can be found, according to a mechanism proposed in []

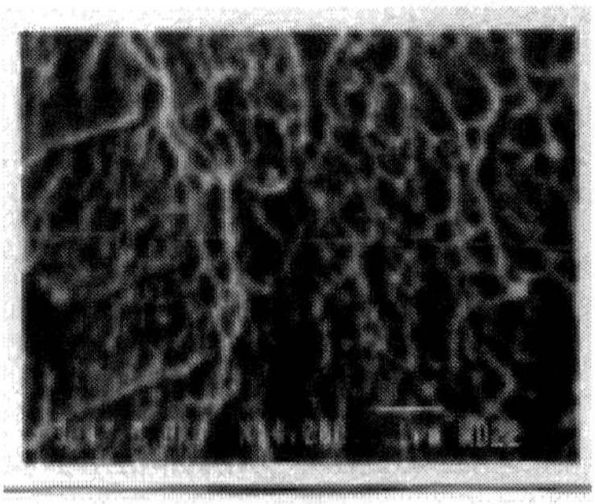

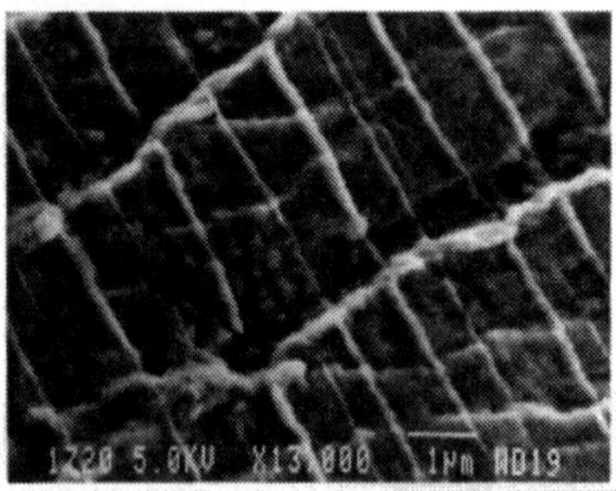

Figure 10a Micro dimples in factes   Figure 10b striations under VA loading

In the case of the variable amplitude spectrum studied here, and for the 8090 alloy, a special kind of striations are formed as seen in figure . The mechanism associated with the formation of these striations has been discussed in [7,8]. It has been associated to a lo-hi type of interaction between low load ratio Ground air ground cycles and high load ratio gust load cycles. It has also been shown that the formation of these striations in this alloy leads to a four fold crack acceleration as compared to what could be expected from constant amplitude behavior. based on these facts, life estimation can be made for the 8090 alloy under the spectrum loading using the procedure shown below.

Considering this crack acceleration factor, a life estimation can be made under the VA conditions studied here.

This estimation was made using the following relationship:

$$ai = ai\text{-}1 + \Delta a \quad (1a)$$

$$\Delta a = da_{ca} * IF \quad (1b)$$

IF is the interaction factor = 3.6

$$da_{ca} = A\, \Delta K^{m} \quad (2)$$

where ai is the current crack length, ai-1 is the previous crack length.

Equation 1b represents accelerated crack growth, only for the high R ratio cycles (not applied for the Ground –Air-Ground cycles)

A and m represent the Paris 'law constants for the Al li alloy []

In table 4, different life estimations made are compared with the observed life.

Table 4 Comparison of life estimations with measure life

Life in no. of flights	experimental	Est1	Est2
	89000	189000	94400

In the first estimation (est1), IF factor is considered to be 1 (linear summation) and in the second one (est2) IF was taken to be 3.6. The first estimation is largely conservative. It can also be seen here that with the second estimation, the estimated life is very close to the experimentally measured life, thus justifying the crack acceleration hypothesis [9]

Conclusions

The fatigue crack growth resistance of aluminum alloys depends upon the precipitate coherency and the environment.

For most of the studied variable amplitude testing conditions here, the 8090 Al Li alloy has the best crack growth resistance.

This advantages is not seen under the variable amplitude loadins spectrum studied here where a load interaction effect leads to the formation of a new kind of striations.

Under such conditions, a crack acceleration is found.

Accurate life predictions can be made if this acceleration effect is taken into account.

References

1 Ranganathan, N., Petit, J. and de Fouquet, J ' *Advances in Fracture Reseasrch* ', Pergamon Press, Volume 3, pp 1767 - 1774.(1984)

2 Bleuzen, C. ,Chaudonneret, M., Farcy,L., Flavenot, J.F. and Ranganathan, N., ASTM STP 1231, American Society for Testing and materials Philadelphia, USA,1994, pp. 504-526 (1994)

3 Ranganathan, N. Thesis, " Contribution au développement d'une approche énreggétique à la fissuration par fatigue", University of Potiers, France (1985)

4 Petit, J., Tintillier, R. Ranganathan, N., Ait-Abdedaim, M. and Chalant, G., 1 '*Fatigue Crack Growth under variable amplitude loading*', Elsevier Applied Sciences Pubs., pp 162-168 (1988)

5 Ranganathan, N., Benguediab, M. ,Nicolas, C., Henaff, G. and Petit, J. Engineering Fracture Mechanics, Vol.42, n°1, pp 59-71.(1992)

6 Petit, .J., 'Fatigue Crack Growth Threshold Concepts ', D.Davidson and S.Suresh eds, The TMS AIME publications, Philadelphia, pp 3-24 (1984)

7 Li, S.Q., Dickson, J.I. and Baïlon, J.P., Materials Sciences and Engineering A, volume 119,pp59-72 (1989)

8 Ranganathan, N. , Li, S.Q., and Baïlon, J.P,, Material Sciences and Engineering, volume 187, pp. 37-42 (1994,).

9 Ranganathan, N. Adiwijayanto, F., Petit, J. and Bailon, J.P., Acta Metallurgica and Materialia, **43**, pp 1029-1035 (1995)

Fatigue of Materials II Advances and Emergences in Understanding
Edited by: T.S. Srivatsan, M. Ashraf Imam, and R. Srinivasan
TMS (The Minerals, Metals & Materials Society), 2013

HIGH FREQUENCY VIBRATION BASED FATIGUE TESTING OF DEVELOPMENTAL ALLOYS

Casey M. Holycross[1,2], Raghavan Srinivasan[2], Tommy J. George[1], Seshacharyulu Tamirisakandala[3], Stephan M. Russ[4]

[1]Air Force Research Laboratory, Propulsion Directorate,
Building 18D, 1950 Fifth Street, Wright-Patterson AFB, OH 45433, USA
[2]Wright State University, Department of Mechanical and Materials Engineering,
3640 Colonel Glenn Highway, Dayton OH 45435, USA
[3]RTI International Metals, Inc.
1000 Warren Avenue, Niles OH 44446, USA
[4]Air Force Research Laboratory, Materials and Manufacturing Directorate,
Building 655, 2230 Tenth Street, Wright-Patterson AFB, OH 45433, USA

Keywords: Fatigue, Titanium, Vibration, Powder-metallurgy

Abstract

Many fatigue test methods have been previously developed to rapidly evaluate fatigue behavior. This increased test speed can come at some expense, since these methods may require non-standard specimen geometry or increased facility and equipment capability. One such method, developed by George et al, involves a base-excited plate specimen driven into a high frequency bending resonant mode. This resonant mode is of sufficient frequency (typically 1200 to 1700 Hertz) to accumulate 10^7 cycles in a few hours. One of the main limitations of this test method is that fatigue cracking is almost certainly guaranteed to be surface initiated at regions of high stress. This brings into question the validity of the fatigue test results, as compared to more traditional uniaxial, smooth-bar testing, since high stresses are subjecting only a small volume to fatigue damage. This limitation also brings into question the suitability of this method to screen developmental alloys, should their initiation life be governed by subsurface flaws. However, if applicable, the rapid generation of fatigue data using this method would facilitate faster design iterations, identifying more quickly, material and manufacturing process deficiencies. The developmental alloy used in this study was a powder metallurgy boron-modified Ti-6Al-4V, a new alloy currently being considered for gas turbine engine fan blades. Plate specimens were subjected to fully reversed bending fatigue. Results are compared with existing data from commercially available Ti-6Al-4V using both vibration based and more traditional fatigue test methods.

Introduction

Fatigue testing, especially for developmental materials, is costly in terms of financial investment and time requirements. An ideal test method would produce accurate, representative data at low cost, and in short order. Generally, data generated using servo-hydraulic test machines is confined to uniaxial stress states, with stress ratios, R, greater than zero. These machines operate on the order of 60 Hz or less, with many hours, days, weeks, or even months, required to fracture a single specimen, generating a single data point.

For the design of many gas turbine engine blades, uniaxial fatigue data at $R>0$ is not a sufficient design parameter, as the predominant mode of failure is high cycle fatigue (HCF)[11] caused by vibration. Typically, the fatigue loading is of a bending or multi-axial nature in a high order bending or torsional vibration mode[8], which cannot be easily simulated using traditional servo-hydraulic test machines. In order to address this discrepancy, a vibration based method was developed by a combined team of researchers from the Air Force Research Laboratory (AFRL) and The Ohio State University (OSU)[8, 16, 9, 15, 7]. This method utilizes an electro-dynamic shaker that applies a base excitation to a cantilevered (fixed-free-free-free boundary condition) square plate, driving it into a high frequency resonant mode (Figure 1). This configuration is more representative of operational fixity and loading conditions of turbine engine blades and can generate fatigue cycles more rapidly than traditional servo-hydraulic load frames. This study also uses an accelerated test technique called the step-test method[10], which further enhances the speed of data generation.

Modes 3 and 4 of a square cantilevered plate are two of the more common mode shape families[4], and as a result, are often of great interest for turbine engine blade fatigue testing[1]. This study focuses on mode 4, commonly referred to as two-stripe mode, for its characteristic stripes of zero displacement, shown in Figure 2. As with displacement, at resonance, there is a characteristic stress field. For this mode, the von Mises stress is shown in Figure 2(b) which has a region of high stress located at the midpoint of the free edge, commonly referred to as the "fatigue zone".

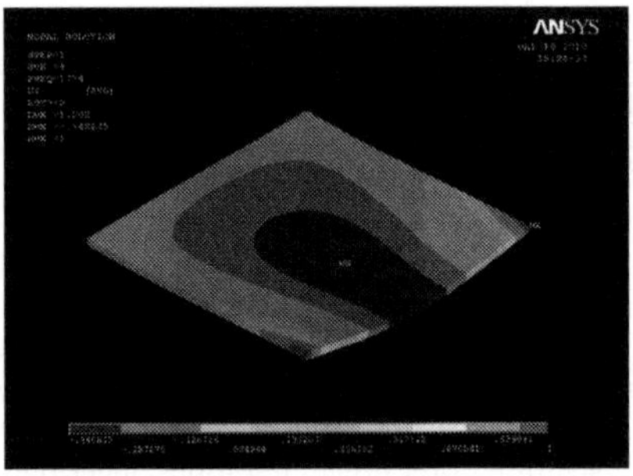

Figure 1: Characteristic shape of mode 4 (1st chord-wise bend, or two-stripe mode) of a square, cantilevered plate

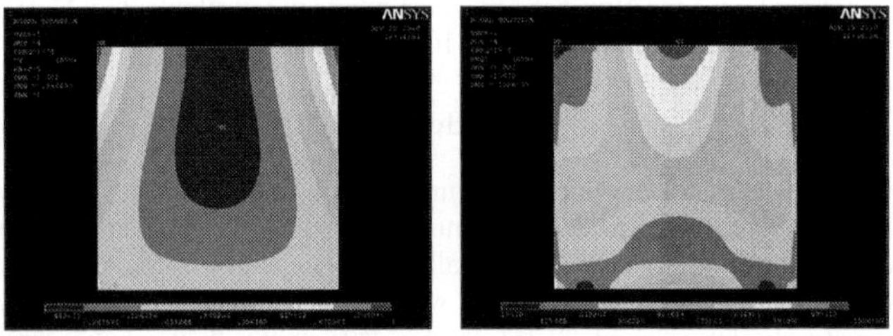

(a) (b)

Figure 2: FEM results for two-stripe mode (a) out-of-plane displacement and (b) von Mises stress. Plate is fixed along lower edge.

Subsurface fatigue crack initiation is unlikely since the fatigue zone is small and the highest stress occurs on the specimen surface and diminishes through the thickness proportional to the distance from the neutral plane. This is a drastic departure from the more common axial fatigue testing procedure, where the fatigue zone, in some cases, can approach the cubic centimeter range. This large volume and uniform stress state allows for subsurface, life limiting flaws to initiate fatigue cracking, giving engineers a useful diagnostic tool to refine material and manufacturing processes. This stark difference in specimen design and test methods will be investigated to determine if fatigue cracks initiate in locations outside of the fatigue zone in the presence of a life limiting flaw.

Material

Boron-modified titanium alloys can be synthesized by a variety of conventional and non-conventional techniques[12] to produce desired properties by controlling size, aspect ratio, alignment, etc. of microstructural phases. More conventional ingot metallurgy techniques generally produce large grain sizes and non-uniform distribution of particulates[18]. This requires additional ingot breakdown and homogenization to obtain the desired microstructure and mechanical properties. Powder metallurgy (P/M) techniques eliminate this need at a significant cost savings. The powder used in this study followed a pre-alloyed powder processing path prepared by induction skull melting and high pressure, inert gas atomization to a nominal composition of Ti-6Al-4V+1B (wt. %). The powder was collected in a cyclone separator and sieved to a nominal size of 149μm. This was done to reduce the amount and/or size of foreign inclusions, a dominant crack initiation mechanism for P/M components. The powder was then degassed, sealed, and subjected to hot isostatic pressing. The ingot was then subjected to a heat treatment and open die forged. The forging was finished to a thickness of 76 mm and diameter of 445 mm.

In order to determine the first tier material properties of the Ti-6-4+1B forging, tensile tests were performed three times in each characteristic direction (axial, radial, and circumferential). Each test was conducted according to ASTM standard E8-09[3]. The results are summarized in Table 1. The yield strengths in each direction were higher than that of Ti-6-4 extra low interstitial (ELI)[5]. The axial direction showed the lowest yield strength (6.2% greater than Ti-6-4 ELI), whereas the radial direction showed the greatest (21.4% greater than Ti-6-4 ELI). The ultimate tensile strength shows similar results, with an increase of 15% and 23% increase over Ti-6-4 ELI for the axial and radial directions, respectively. Elongation suffered a debit in the axial direction. However, the circumferential and radial directions exhibited slightly higher elongation. The stiffness was greatest in the circumferential direction, with an increase of 20.4%, whereas the lowest stiffness occurred in the axial direction, only increasing by 7%. This is consistent with the tensile results of rolled plate of the same composition[13].

Methods and Theory

From the forging, specimens were wire electrodischarge machined to a nominal size of 177.8×114.3×3.18 mm. Testing was conducted using a Ling 18,000 lbf electrodynamic shaker outfitted with a clamp, making the nominal effective test section of the specimen 114.3×114.3×3.18 mm. The specimen was excited by a forced vibration at its two-stripe modal frequency creating a fully reversed bending stress at the midpoint of the free edge[8].

The specimen was instrumented with a strain gage at the location of maximum stress and a laser vibrometer target (Figure 3). At high stresses and long test times, the strain gage failed internally or delaminated from the specimen rendering it useless. In order to continue testing, a reference displacement at the laser vibrometer target location was calibrated to the strain response. The calibration was performed by acquiring data from the strain gage and the laser vibrometer at increasing shaker drive input. A least squares regression line of the relationship between displacement and strain was calculated, and strains above which the gage fails were extrapolated from the measured displacement. Clamp fixity and small geometric variations can play a significant role in response, so a calibration curve was generated for each specimen prior to fatigue testing.

The fatigue tests were conducted using an accelerated test technique developed by Maxwell and Nicholas[10]. This method begins by selecting a starting stress well below the expected fatigue strength. This stress was applied for 10^6 cycles, and if failure did not occur, the stress was incremented or "stepped" up, and the cycling is repeated until failure. The failure stress is calculated using Equation 1.

$$\sigma_{\text{fail}} = \sigma_{\text{pr}} + \left(\frac{N_{\text{fail}}}{N_{\text{step}}}\right) \Delta\sigma \tag{1}$$

where σ_{fail} is the estimated failure stress, σ_{pr} is the stress level from last step in which failure did not occur, N_{fail} is the number of cycles completed at the stress level of failure, N_{step} is the number of cycles in a complete step (10^6 for this study), and $\Delta\sigma$ is the step size.

Table 1: Tensile test results[17]

Direction	Yield Stress (MPa)	Ultimate Tensile Stress (MPa)	Elongation (%)	Elastic Modulus (GPa)
Axial	842	991	5.0	122.0
Circumferential	926	1020	13	137.2
Radial	963	1054	12	129.5
Ti-6-4 ELI	793	862	10	114

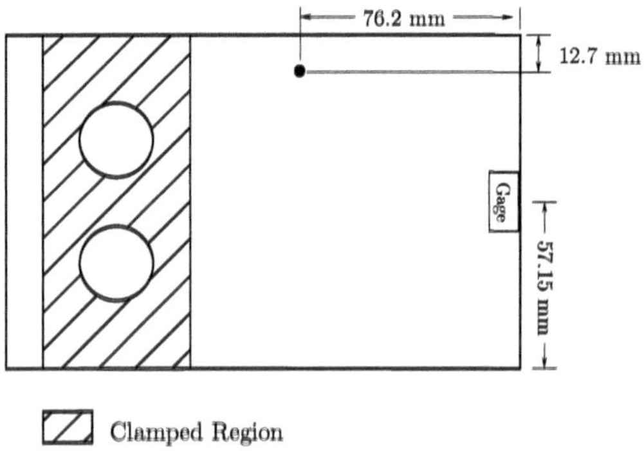

Figure 3: Velocity target (•) and strain gage location

According to Maxwell and Nicholas, step sizes of 10% or less of the expected fatigue stress is preferred[10]. Since Ti-6-4+1B is considered a developmental alloy that was previously shown to have life limiting flaws[13], the beginning stress level was initially chosen as 172 MPa (25 ksi) with a step size of 34.5 MPa (5 ksi). This was sufficiently low to ensure that a full step was completed before failure.

Results

The 10^6 fatigue strengths for the Ti-6-4+1B specimens are presented Figure 4 as calculated by Equation 1 along with comparative data for wrought[14] and P/M[6] Ti-6-4. The mean fatigue strength from nine tests of the Ti-6-4+1B forging was 332 MPa (48.1 ksi) with a range from 207 to 457 MPa (30 to 66.3 ksi). The mean fatigue strength was 37.9% lower than the minimum strength for wrought Ti-6-4 using the same test method[14] and is inconsistent with previous studies of both boron modified P/M Ti-6-4[13]. Fractographic analysis was employed to gain additional insight into the cause of these poor properties.

The initiation sites of the fatigue cracks were easily identified visually by locating the center of concentric beach marks or the convergence of river lines to a single point (Figure 5(a)). Further inspected using electron microscopy, Figure 5 identified that this region for specimen T3 had an agglomeration of artifacts near the plate surface. Figure 5(c) indicates several locations where energy dispersive x-ray spectroscopy (EDS) analysis was completed. The EDS results are provided in Table 2, indicating several locations rich in carbon, sodium, silicon, sulfur, chlorine, potassium, calcium, and iron, suggesting intermetallic inclusions.

Figure 6 shows the location of the crack initiation sites for each plate with von Mises stress contours from FEM. All initiation sites are bounded by the first three contour lines,

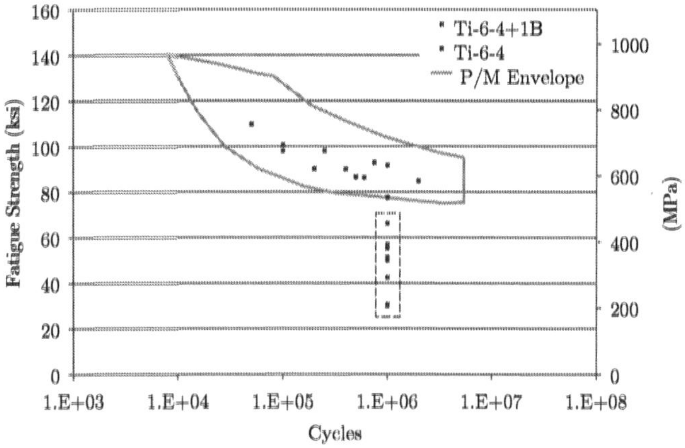

Figure 4: Experimental fatigue data comparison to previous Ti-6-4 data using the same test method as this study[14] and the typical fatigue performance envelope for P/M Ti-6-4[6]

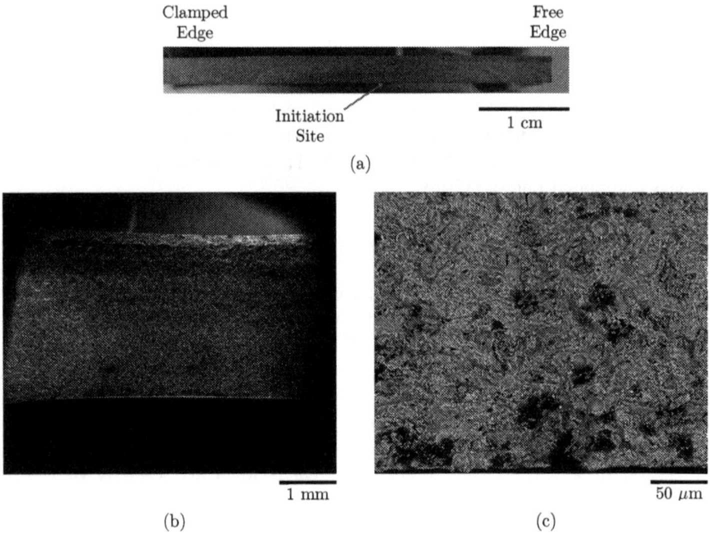

Figure 5: Secondary electron micrograph (b) of plate T3 showing characteristic river lines and backscatter electron micrograph (c) of the fatigue initiation site

Table 2: EDS results for plate T3

Location	Elements detected
0	Ti, Al
1	Ti, V, Al, O
2	Ti, V, Al, O, C, Na, Si, S, Cl, K, Ca, Fe
3	Ti, V, Al, O, C, Na, Si, S, Cl, Ca, Fe
4	Ti, V, Al, O, C, Si, S, K, Ca, Fe
5	Ti, V, Al, O, Si, S
6	Ti, V, Al, O, C, Na, Si, S, Cl, K, Ca, Fe
7	Ti, V, Al, O, C, Si, Fe
8	Ti, V, Al, O, C, Si, S, Fe
9	Ti, V

which represent, starting from the midpoint of the free edge, 94.5%, 83.4%, and 72.4% of maximum stress.

Discussion

The 10^6 cycle fatigue performance of the Ti-6-4+1B was poor in comparison to Ti-6-4 data generated using the same test method. Furthermore, the fatigue strengths had a high degree of variability, of which, none fell within the typical fatigue strength scatter envelope

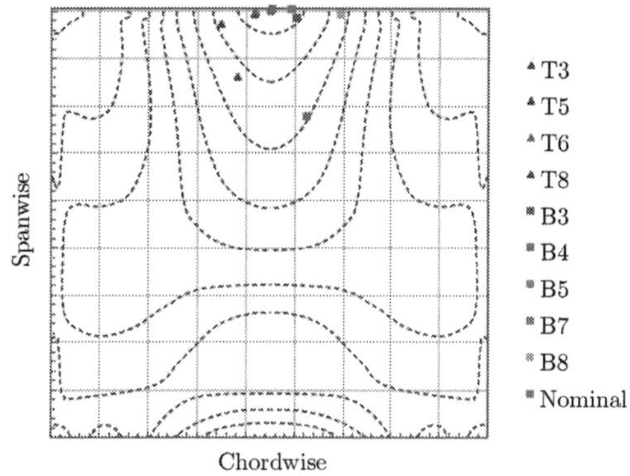

Figure 6: Crack initiation sites overlaid with von Mises stress contours. The plate is fixed along the lower edge.

for pre-alloyed powder metallurgy Ti-6-4. Fatigue cracks initiated at the surface for all test plates, with all initiation sites being suspect of intermetallic inclusions. The vibration-based testing method, despite having a relatively small fatigue zone, still exhibited preferential failure in life limiting inclusion regions, demonstrating its practicality for the fatigue testing of developmental alloys. The total time required to complete the fatigue testing was roughly seven hours. Testing using servo-hydraulic load frames operating at 60 Hz would have required about 170 hours. Test time using this vibration based method, reduced by more than an order of magnitude, affords a more rapid fatigue assessment of developmental alloys and accelerated feedback of the presence of potentially life-limiting defects to the alloy development process.

Acknowledgments

The authors would like to thank the Propulsion and the Materials and Manufacturing Directorates of AFRL for financial, facility, and equipment support.

References

[1] Kyle S. Allen. Evaluation Techniques for Determining Damping Mechanisms on Titanium Plates. Master's thesis, Air Force Institute of Technology, WPAFB, 2005.

[2] ASTM International. *ASTM E3-01: Standard Guide for Preparation of Metallographic Specimens.* 2007.

[3] ASTM International. *ASTM E8-09: Standard Test Methods for Tension Testing of Metallic Materials.* 2009.

[4] Christopher M. Blackwell. Evaluation of the Damping Characteristics of a Hard Coating on Titanium. Master's thesis, Air Force Institute of Technology, 2004.

[5] M. J. Donachie Jr. *Titanium: A Technical Guide.* American Society for Metals, Metals Park, Ohio, 2000.

[6] F. H. Froes. Developments in Titanium P/M. *Electronic resource,* (www.webs1.uidaho.edu/imap/MPR%20Paper.pdf), 2010.

[7] Tommy George. *Development of methodologies for ensuring structural safety of gas tubines and launch vehicles.* PhD thesis, The Ohio State University, 2002.

[8] Tommy George, Jeremy Seidt, M.-H. H. Shen, Theodore Nicholas, and Charles Cross. Development of a Novel Vibration-based Fatigue Testing Methodology. *International Journal of Fatigue,* 26:477–486, 2004.

[9] Tommy J. George, M.-H. H. Shen, Onome Scott-Emuakpor, Theodore Nicholas, Charles J. Cross, and Jeffrey Calcaterra. Goodman Diagram via Vibration-based Fatigue Testing. *Journal of Engineering Materials and Technology,* 127:58–64, January 2005.

[10] D. C. Maxwell and T. Nicholas. A Rapid Method for Generation of a Haigh Diagram for High Cycle Fatigue. *Fatigue and Fracture Mechanics: 29th Volume, ASTM STP 1332,* pages 626–641, 1999.

[11] T. Nicholas. Critical Issues in High Cycle Fatigue. *International Journal of Fatigue,* 21:S221–231, 1999.

[12] K. B. Panda and K. Ravichandran. Synthesis of Ductile Titanium-Titanium Boride (Ti-TiB) Composites with a Beta-Titanium Matrix: The Nature of TiB Formation and Composite Properties. *Metallurgical and Materials Transactions A,* 34:1371–1385, June 2003.

[13] Kevin A. Schwendiman. Critical Life Prediction Research on Boron-Enhanced Ti-6Al-4V. Master's thesis, Air Force Institute of Technology, 2007.

[14] Onome Scott-Emuakpor. *Development of a Novel Energy-Based Method for Multi-axial Fatigue Strength Assessment.* PhD thesis, Ohio State University, 2007.

[15] Jeremy Seidt. Development of a novel vibration based high cycle fatigue test method. Master's thesis, The Ohio State University, 2001.

[16] M.-H. H. Shen, J. Seidt, C. Cross, P. W. Whaley, and T. Nicholas. Development of a Novel Method for Evaluating Material Behavior Under Turbine Engine Operating Conditions, Part II: An Empirical Vibration-Based Fatigue Assessment Framework. In *6th National Turbine Engine High Cycle Fatigue Conference,* 2001.

[17] United States Air Force. FA 8650-08-D-2806: VAATE Titanium Matrix Composite FLADE Demonstration.

[18] Fred Yolton. The Pre-alloyed Powder Metallurgy of Titanium with Boron and Carbon Additions. *Journal of the Minerals, Metals and Materials Society,* pages 56–59, May 2003.

OVERVIEW II

Fatigue of Materials II Advances and Emergences in Understanding
Edited by: T.S. Srivatsan, M. Ashraf Imam, and R. Srinivasan
TMS (The Minerals, Metals & Materials Society), 2013

DWELL FATIGUE DESIGN CRITERIA

Tarun Goswami

Damage Tolerance and Probabilistic Life Prediction of Materials Center
College of Engineering and Computer Science
Wright State University
3640 Colonel Glenn Hwy
Dayton, Ohio 45435, U.S.A

Abstract

Many engineering components operate under steady state conditions for a period of time after the peak loads had been achieved. The time during which the loads are near constant is considered dwell time. Materials exhibit sensitivity to dwell times differently to the tension and compression loads and continue to challenge engineers designing such systems. New dwell sensitivity maps proposed in this paper showing the dwell sensitivity behavior of materials in low cycle fatigue (LCF) where life is determined by the development of plastic/inelastic strains and long crack growth (LCG) which is also known as high cycle fatigue life of components. While dwell sensitivity in LCF was significant in lower ranges of strains, higher strains and dwell time did not lower fatigue life as highly as it was under lower strains. LCF dwell sensitivity was predicted with the use of strength ratio, dynamic balance between the hardening and softening, and dwell time damage parameter. Each of these characteristic features predicted dwell sensitivity.

Dwell sensitivity in LCG has not been investigated as widely as in LCF. A new parameter proposed which measures the sensitivity in LCG called, normalized crack growth ratio (NCGR). It was a ratio of crack growth rates under continuous fatigue and dwell fatigue at the same mode I stress intensity factor range. The LCG dwell sensitivity was found from an order-of-magnitude to four orders-of-magnitude. Therefore, the dwell sensitivity was more significant in LCG. Dwell fatigue crack growth mechanism map was generated for Inconel 718 charting various zones in which a particular damage mode was observed. Remaining life assessment performed for a disk component with dwell fatigue crack growth models showed negligible life extension potential under the assumed conditions. There is a need for further collaboration and consensus development defining and/or refining the input parameters.

Keywords: fatigue, long crack growth, low cycle fatigue, crack, dwell sensitivity.

Introduction

The operating conditions of many engineering components require service loads to change with respect to time to meet a specific demand. The start-up and shutdown cycles impose low cycle fatigue damage at localized areas, where the material is inhomogeneous or discontinuous. In these localized areas, the stress is concentrated and amplified due to geometrical constraints, inherent anomalies, induced anomalies due to handling and service usage and micro-cracks that may generate with respect to usage. A major part of application requires that a component be held at near steady conditions, such as during the period between the shut downs of a gas turbine engine or power plant turbine disk. The time during which load and/or stress is steady, is known as dwell or hold time. During a dwell time, if appropriate conditions met, creep deformation takes place. These conditions are high temperature, steady stress and time. Therefore, failure mechanisms under such loading conditions are a result of complex interactions of creep and fatigue processes, and/or environment, which result in premature failure within low cycle regime typically ranging from 10-10,000 cycles. For gas turbine blades and discs, life may range in several thousand hours. Fewer components require the opposite, where loads held constant under peak compression loads, for example, rolls in hot milling application or dies in extrusion mills. Therefore, dwell effects and creep-fatigue life prediction issues are very important in design in order to understand the failure mechanisms and life assessment of components.

The creep-fatigue failure modes or low cycle fatigue (LCF) failures documented in components from power and aviation industry from early 1950's. While commercial aviation engines experienced disc bursts due to dwell effects [1-2], recent disc bursts on forged titanium alloy discs [3-4] became a major cause of concern needing further research. Premature disc bursts also occurred in power plants. Hence, dwell cycles, during which creep and fatigue processes interact and lower the high temperature fatigue resistance of materials is very important requiring further studies. Much of the efforts [5-44] focused on defining the conditions that result in LCF failure, there is no consensus in the community on the appropriate crack size definitions (both initial and final) that would establish the life assessment protocols in LCF. Experience has shown that the component life is comprised of at least two phases, 1) LCF where anomalies and/or crack present transition to LEFM crack, and 2) long crack propagation to failure conditions.

The fatigue design criteria separates the useful life in: 1) nucleation of a crack, 2) propagation within the first few grains, 3) propagation below the threshold stress intensity factor range for long cracks and subsequent transition of the short crack or a defect to long crack and, 4) crack propagation of long crack to a safe critical size, defined as failure. The disk component considered in this study, used in gas turbines, designed based on safe life and retired upon reaching the low cycle fatigue limit. Crack length at the end of LCF life considered to be from 0.025-0.075 mm. However, this micro-crack formation does not take into account the discontinuities present, service induced defects and inherent defects due to material manufacturing. The low cycle fatigue life varies in the presence of defect(s) and therefore, there is a need to develop fatigue design criteria that not only account for dwell but also inherent and induced defects.

While experience has shown that start-up and shutdown conditions represented by strain controlled LCF tests, the period between the shut down relates centrifugal loads and represented by high cycle fatigue where crack propagates assuming linear elastic fracture mechanics considerations. These mechanisms are very difficult to separate in lack of appropriate methodologies. An attempt was made to summarize the dwell effects in LCF life and fatigue crack propagation rates in candidate materials in this paper. Numerous dwell cycles are possible, such as tensile dwell (t/0), compressive dwell (0/t), equal dwell in tension and compression (t/t, balanced dwell cycle) and unequal dwell cycles (t_1/t_2, unbalanced dwell cycles). The materials are sensitive to dwell cycles differently. Some materials exhibit sensitivity to tensile dwell and others to compression dwell as well as rates at which deformation applied, for example, high and low rates. Therefore, there is a need to understand the dwell fatigue behavior of materials for different engineering applications. Damage development under dwell cycles is different as well for tension and compression dwells. Figure 1 shows the relations between the strain-time, stress-time and stress-strain behaviors under continuous fatigue, tension dwell, compression dwell, and tension and compression dwell cycles.

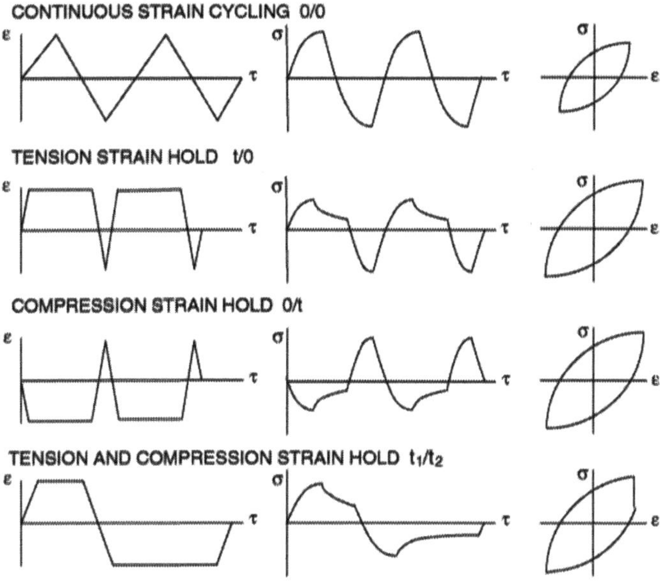

Figure 1: Examples of fatigue waveforms used in laboratory testing [45]

Therefore, the objective of this paper is to investigate the dwell sensitivity in LCF and in long crack propagation, generating material behaviors to perform the analytical life assessment, and compare the sensitivity of the disk with non-dwell data of a turbine engine disk.

DWELL SENSITIVITY BEHAVIOR (LCF)

Previous work of the author shows the effects of dwell times on the elevated temperature LCF behavior of high temperature materials. A dwell sensitivity map proposed to show the beneficial and/or detrimental effects of dwell cycles with respect to continuous fatigue behavior. The map plots the plastic strain range and normalized cycle ratio (NCR), at a given temperature. When NCR was less than one, material was dwell sensitive to that type of cycle. However, when NCR was higher than one, dwell cycle caused beneficial effects, thereby improving the HTLCF resistance. Continuous fatigue life taken as a reference to measure dwell sensitivity. Two groups of tests were observed in the literature, 1) isothermal fatigue (IF), Figure 2, and 2) thermo-mechanical fatigue (TMF), Figure 3, and used to derive NCR values and dwell sensitivity maps. In the former category deformation rate was also used to characterize the LCF behavior, however, in the latter category in-phase cycles were used which apply maximum strain and temperature at the same time.

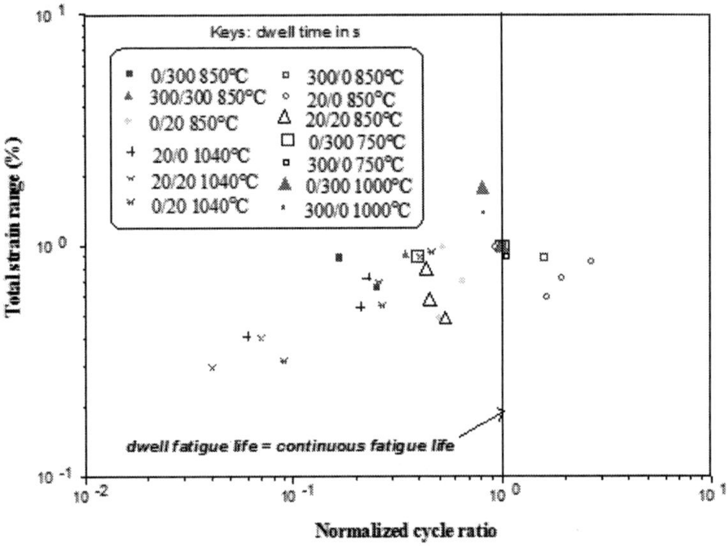

Figure 2: Dwell sensitivity behavior in isothermal, low cycle fatigue [46]

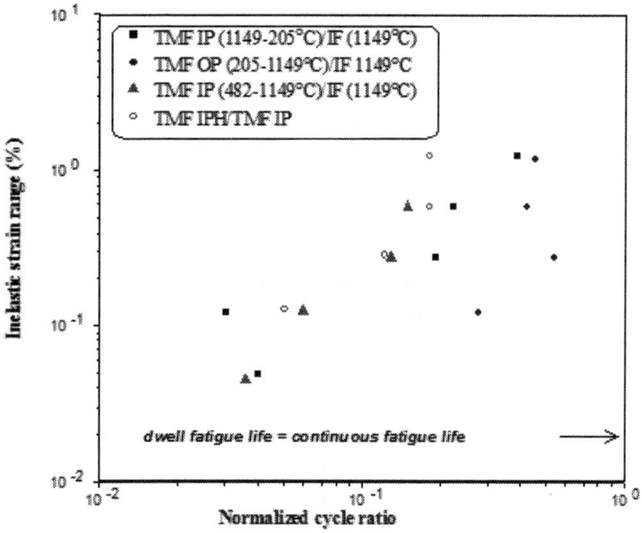

Figure 3: Dwell sensitivity behavior in thermal fatigue [46]

The LCF tests conducted using cylindrical specimens under displacement control. Since higher strain ranges applied in the accelerated laboratory testing, the specimens concentrate high strain on the gage length of the specimens. From these localized high strain regions cracks form. The number of cracks that forms on the specimen surface is a function of test temperature, strain range and dwell time parameters used in testing. Apart from these test parameters environment plays a major role in the interaction mechanisms by compounding the damage by oxidation mechanisms. In one select case, LCF testing of IMI 829, total number of cracks formed is shown in Figure 4. It demonstrates that number of cracks that form on the surface is directly proportional to total strain range applied. While compressive dwells were more deleterious for IMI 829 at high temperature, the interaction mechanisms included not only the fatigue and creep deformation but also environmental oxidation.

The dwell sensitivity maps proposed here may be applied in design and life assessment by extrapolating data for longer dwell periods where such data are not available. Only a summary of dwell sensitivity behavior was presented below. Publications [45-50] present the material behaviors under individual test conditions, NCR-strain range distributions, and other relations more fully as well as other uses of [5-44]. The following dwell sensitivity behaviors are noteworthy:

- All the materials studied were dwell sensitive at lower strain ranges in this research and therefore, NCR values were low below 1.0% total strain range.
- At high strain ranges (above 1.0% total strain range), saturation in the dwell sensitivity occurred and as a result NCR values increased. Saturation in this context meant further increase in strain range did not lower life further.
- Dwell sensitivity and temperature sensitivity interacted for two materials, namely SS 316 and Ni 201 [49]. Temperature sensitivity resulted in higher IG cracking, loss in ductility, and the mode in which the deformation accumulated (for example in SS 316 at 600°C by cell type, whereas at 700°C by means of sub-boundary type dislocation sub-structure).
- A particular combination of test parameters, for example, temperature, dwell time and strain rate, improved the creep-fatigue resistance of materials. Formation of new precipitates was beneficial as they blocked the grain boundary sliding and IG cracking by creep. The NCR values in those conditions increased and became close to 1 or higher.

The concept presented above and in the figure below, it is evident that number of cracks that form on the surface of the disk will be a function of higher strain range or associated plastic strains. Since disk components operate below yield strength in elastic strain dominated region, the number of cracks from service exposure will be limited and assumption that only one crack may be present at the end of LCF life may be valid, though requires more research and characterization of previously tested specimens. This is one critical area, which requires the attention of the lifing community to develop appropriate consensus while accepting more risk yet optimizing the economic reasons to use further the component.

Damage Mechanisms (LCF Dwell)

Damage mechanisms under dwell cycles were different for different materials. Some materials found to be cyclic softening, while others were cyclic hardening type with the application of continuous fatigue and/or dwell cycles. Deformation in a material was also material dependent, some materials exhibited intergranular (IG) damage and others transgranular (TG) damage. However, mixed damage found in long crack growth at high temperature with dwells. In some materials oxidation damage occurred from exposure to temperature and time, which produced surface damage. Precipitates occurred and concentrated along the grain boundaries and caused IG damage. Tensile dwells caused a shift of stress in the compression direction, thereby improving the HTLCF resistance in some materials. However, for other cases such as 1Cr-Mo-V steel, the tensile dwell caused cavitation or creep damage, resulting in stress concentration in tension direction and caused tensile dwell sensitivity.

The mechanisms controlling deformation and failure summarized below:

- The dwell sensitivity was dependent upon the materials for example, microstructure, condition and grain size. Only after a particular temperature was exceeded did the deformation in a material accrue by means of precipitation of carbides. Many metal

(M) carbides ($M_{23}C_6$) can form along the grain boundaries. These were of Cr, Mo, C and other alloying elements. As these precipitates formed, slip systems became more mobile and concentrated along the carbides, or the grain boundaries that promoted grain boundary sliding, cavity formation and IG damage causing dwell sensitivity.

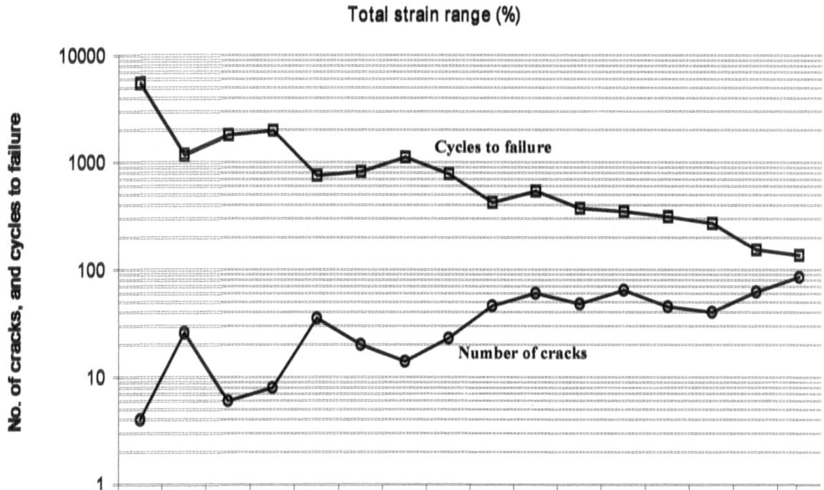

	1	1	1	1	1	1	1.5	1.5	1.5	1.5	1.5	1.5	1.5	2.5	2.5	2.5
—●—3	4	26	6	8	35	20	14	23	46	60	48	64	45	40	62	85
—■—3355	5500	1182	1800	1963	755	819	1115	781	423	536	371	349	311	271	153	136

Dwell time 0/0 2/0 0/2 2/2 15/0 0/15 15/15 0/0 2/0 0/2 2/2 15/0 0/15 15/15 0/0 15/0 15/15
min

Figure 4: Number of secondary cracks with respect to strain range and dwell time in IMI 829 tested at 600°C. The ordinate shows number of cracks and cycles to failure, whereas abscissa shows total strain range (%). Note: first row: percent total strain range, second row: number of cracks by open circle, third row: cycles to failure by open square. Number of cracks peaked for compressive dwell periods and reduced cycles to failure thus making this material a compressive dwell sensitive material.

- The damage also was dependent upon the test temperature, strain rate, dwell time and direction in which it was applied. With the application of tensile dwell, stress relaxation occurred which lowered the materials' strength in that loading direction. Softening occurred with a combination of lower strain rates, longer dwell periods, lower stress range, and large relaxation strains (which is a result of strain transformation from elastic to plastic) that promoted IG damage. Other features such as cavity formation and grain boundary sliding compounded the damage and failure in this mode was faster than TG mode. This mechanism raised the NCR values at higher strain levels as seen in all materials. This feature may play a key role in Inconel 718 by converting elastic strains to plastic strains under tension dwell cycles and may be more deleterious in that direction.

- Within a range of temperature, mechanical properties of some materials improved for example, 2.25Cr-Mo, Ni 201, and SS316. During this temperature range clusters of carbides formed for example Mo-C due to dynamic strain aging. This action enhanced the fatigue as well as tensile properties of materials within that temperature range. However, above this temperature, more damage occurred as other processes such as environmental interactions accelerated the growth of the damage.
- Oxides formed on the surface of specimen due to exposure at high temperatures. Surface striated damage and crack intrusions formed by either IG/TG or mixed conditions. As the strain range increased the number of such oxide intruded cracks increased. Longer dwell periods caused several cracks to grow within the material inside the gauge area.
- Dwell cycles produced a change in the mechanical properties of materials, where a dwell cycle either lowered the modulus or enhanced it. In most cases as the test temperature increased, the tensile strength properties decreased. With the application of a dwell cycle stress range in both directions increased or decreased. An initial 5-15% of the test, life was in achieving the stable hysteresis loops. Deformation in the material was in three mode, which separates the boundaries where TG, mixed and IG fracture occur. These regions depend upon the stress range applied in a cycle and material softening and hardening phenomenon. Only after exceeding a critical stress, every cycle produced the damage in one of the following three modes TG, mixed and IG. The range of stress where it occurred was not determined for each material studied in this investigation. For some low alloy steels, 1Cr-Mo-V, these details were mapped, however, most other materials are not investigated. It is imperative that lower stress and strain rates produce IG damage where grain boundary sliding and cavity formation occur, whereas, in the other ranges, mixed mode and TG fracture occur with defined striation patterns. For a hardening/softening material, these boundaries were schematically drawn elsewhere [49].
- Trends in the deformation mechanisms were material dependent. Three distinct zones identified where a particular damage occurs. Even-though it is very difficult to define the boundaries in stress range where a particular model should prevail, this study identifies such a possibility.
- For some materials, with the application of tensile dwell, mean stresses developed in tension direction. Positive mean stresses produced creep damage and reduced life. However, for other materials, compressive dwell cycles generated complex environmental interactions of creep, fatigue, and oxidation processes and lowered the life more than an equal tensile dwell cycle.
- An oxidation based damage development model was presented in Publication VII based on the damage mechanistic features observed in the metallography and fractography of two alloys, namely IMI 829 and MAR M 002. The features observed in these materials were complex interactions of creep-fatigue-oxidation mechanisms, which cannot be included in a simple mathematical equation. Number of secondary cracks increased as the strain range and dwell time increased in IMI 829 indicating

surface roughness created by oxidation. Mixed IG/TG cracking, wedge type of cracking, surface striated cracking, oxide filled cracks, and depletion of elements were documented [49].

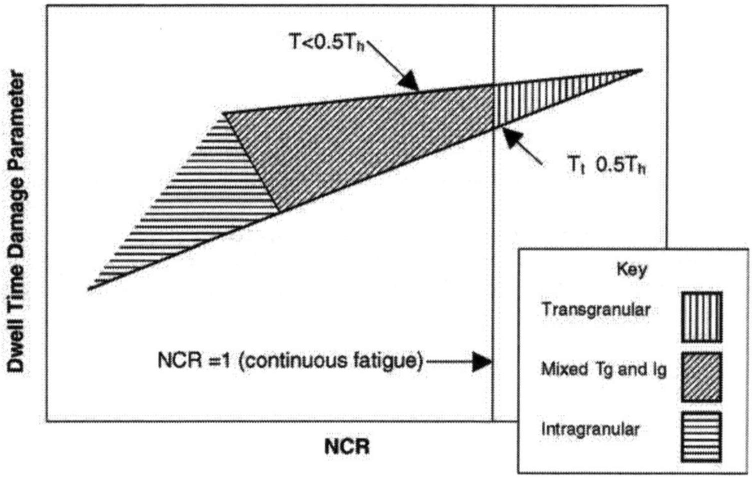

Figure 5: Damage development in dwell fatigue testing at high temperatures, Note (T_h= homologus temperature) [45]

Dwell Sensitivity Criteria (LCF)

From tensile properties: while prediction of dwell sensitivity behavior is very important for design purposes, no attempts made in the literature to predict dwell sensitivity. Attempts made by the author included tensile properties of the materials at room and test temperatures and their ratios. It was found that equal strength ratios (yield to tensile) at room and high temperature predicted compressive dwell sensitive behavior, whereas the range where tensile dwell sensitivity occurred was quite wide under and over the ratio 1. Using the above criteria, dwell sensitivity was predicted for a number of materials.

From dynamic properties: a conceptual model uses material hardening and softening behaviors separately in the tension and compression directions. One of the following four combinations for every creep-fatigue cycle exists in LCF tests:

1) Tensile softening + compressive softening (SS)
2) Tensile softening + compressive hardening (SH)
3) Tensile hardening + compressive softening (HS)
4) Tensile hardening + compressive hardening (HH)

These combinations were determined for a tensile and compressive dwell sensitive material. The dwell sensitivity behavior of materials studied was considerable at lower levels of strain ranges. However, as the strain range increased the sensitivity to dwell times decreased. A combination of test parameters such as dwell time, test temperature, and strain range may provide a threshold, conceptually, where the material is likely to be most sensitive to dwell times. In certain cases dwell times enhanced the creep-fatigue life (for example, copper alloys AMZIRC and NARaloy-Z, and superalloys PWA1480 and MA754), whereas in most other materials, dwell times generally reduced the life.

Dwell time damage parameter: dwell time damage parameter (DTDP) was proposed to represents dwell sensitivity. The DTDP in a product form consists of following fatigue test parameters; relaxed stress, dwell time, strain rate and inelastic strain range with respect to the cycles to failure. The relaxed stress was further separated into tension only relaxed stress for tension dwells, for compression only relaxed stress for compression only dwells, stress range in the direction of dwell, respectively. Equations below show the partitioned relaxed stress components. Using the relaxed stress in the direction of dwell time and cycles to failure correlate the damage in a power law and proposed as one of the dwell sensitivity criterion.

$$\mathrm{DTDP} = \sigma_r \cdot t_\mathrm{h} \cdot \dot{\varepsilon} \cdot \Delta\varepsilon_\mathrm{in},$$

$$\mathrm{DTDP} = \sigma_\mathrm{sdh} \cdot t_\mathrm{h} \cdot \dot{\varepsilon} \cdot \Delta\varepsilon_\mathrm{in},$$

$$\mathrm{DTDP} = \sigma_\mathrm{tsr} \cdot t_\mathrm{h} \cdot \dot{\varepsilon} \cdot \Delta\varepsilon_\mathrm{in},$$

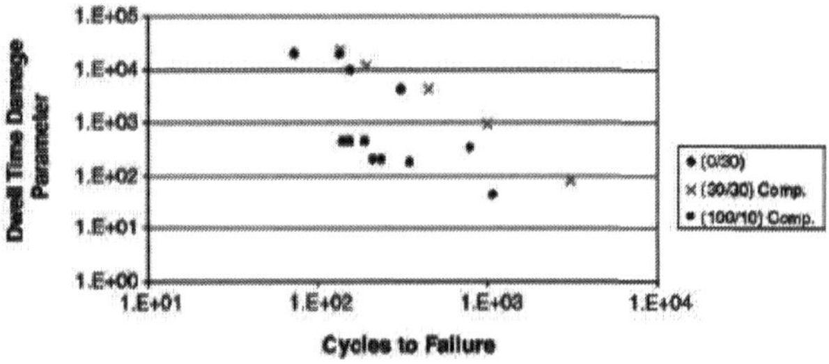

Figure 6: Dwell time damage parameter [45]

DWELL SENSITIVITY IN LONG CRACK GROWTH

The fatigue crack growth behavior of Inconel 718 was compiled from the literature [51-60]. The inclusion criteria included continuous fatigue crack growth at room and high temperature, stress ratio, air and vacuum environment, grain size, condition such as bar, disk and necklace stocks were included, and dwell time during mode I stress intensity factor range. The data compiled during this effort is currently analyzed and prediction model under development [61]. A wide scatter was observed in the data for both categories- continuous fatigue and dwell crack growth as presented in Figure 7 and 8, respectively.

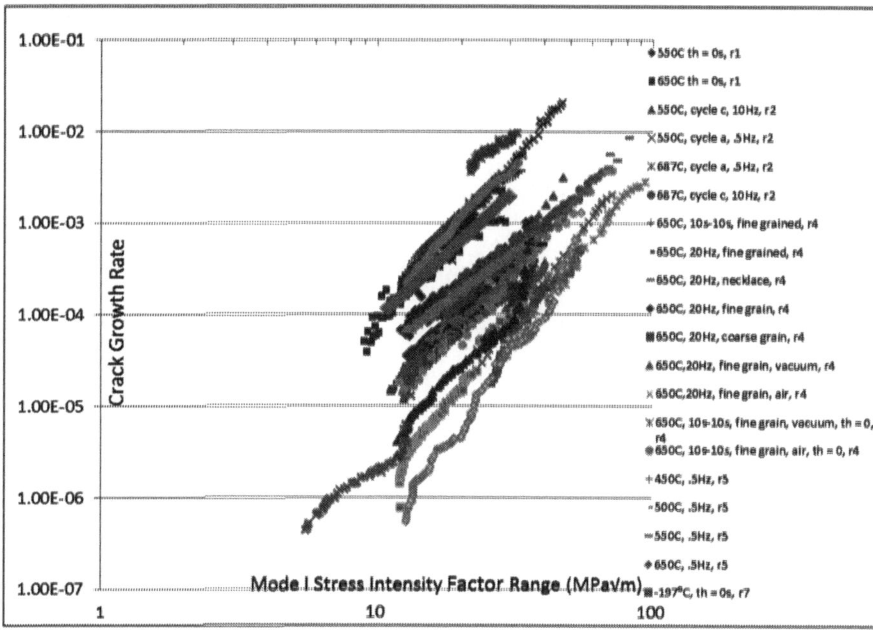

Figure 7: Fatigure crack growth (rate in mm/cycle) behavior in Inonel 718 at differnet temperatures, test frequency, grain size, and stress ratio [51-60].

As evident from Figure 7 that a wide scatter in the material behavior exists for Inconel 718 tested at room and high temperatures. The crack growth rate varies three orders of magnitudes or more under continuous fatigue and demonstrates the need of probabilistics in design.

Dwell Fatigue Crack Growth

Figure 8 shows the dwell fatigue crack growth behavior in Inconel 718 at various testing conditions. Though dwell times varied from 10s to 21,600s, and resulting crack growth rates scatter was 4 orders of magnitudes.

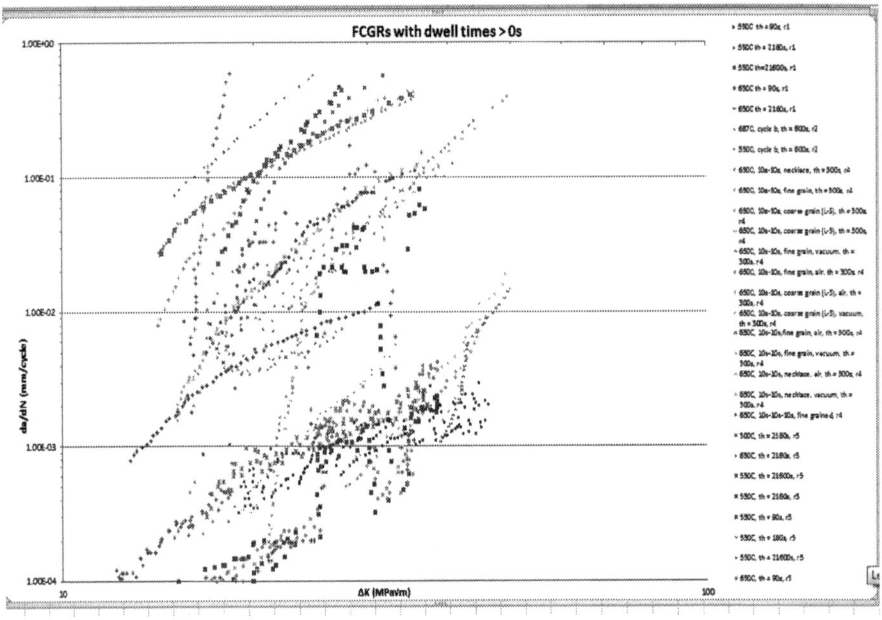

Figure 8: The dwell time fatigue crack growth in Inconel 718 at room and high temperature [51-60]

Dwell Sensitivity Maps in Long Crack Growth

New dwell sensitivity map was proposed. It applies to long crack growth scenarios. A normalized crack growth rate ratio (NCGR), which is a ratio of crack growth rates under continuous fatigue and dwell cycles at the same Mode I stress intensity factor range was evoked and plotted with respect to ΔK. The NCGR compares the dwell fatigue crack growth rates with respect to continuous fatigue and shows that dwell sensitivity ranges from 10-100, for most part, at the low ranges of ΔK, however, as ΔK increases and high dwell times, the dwell sensitivity accelerates. Table 1 shows the salient features of the data used to develop the dwell sensitivity maps in long crack growth regime. Only forged and necklace conditions were included to derive the trends; from a total of 11 conditions, 3 were necklace structures. The temperature cases were 550 and 650ºC and dwell times varied from 10 to 21,600s. All the data were averaged and fitted with a power-law (Paris) equation, the parameters of the equations C and m identified for continuous fatigue and dwell fatigue. At given range of ΔK crack growth rates were determined for both dwell and no dwell, NCGR determined and plotted in Figure 9. The data presented in Fig. 9 shows that smaller dwell time cycles, for example, 10s to 90s at 550ºC show dwell sensitivity at low ranges of ΔK, however, as ΔK increases the material exhibits crack growth resistance.

Table 1: Salient features of the data analyzed for dwell sensitivity

Category	Necklace/Forge	Temp°C	Hold Time (s)	Number Cases	C_avg	m_avg
1	F	650	0	7	3.89E-07	2.37
2	N	650	0	1	2E-08	2.66
3	N	650	300	1	4E-14	7.95
4	F	650	300	7	1.3E-05	3.63
5	F	650	10	1	6E-07	2.94
6	F	650	90	1	5E-06	2.73
7	F	550	0	1	3E-07	2.11
8	F	550	90	1	1E-06	1.95
9	F	550	2160	1	3E-08	4.23
10	F	550	21600	1	2E-11	7.21
11	N	550	650	1	3E-07	2.38

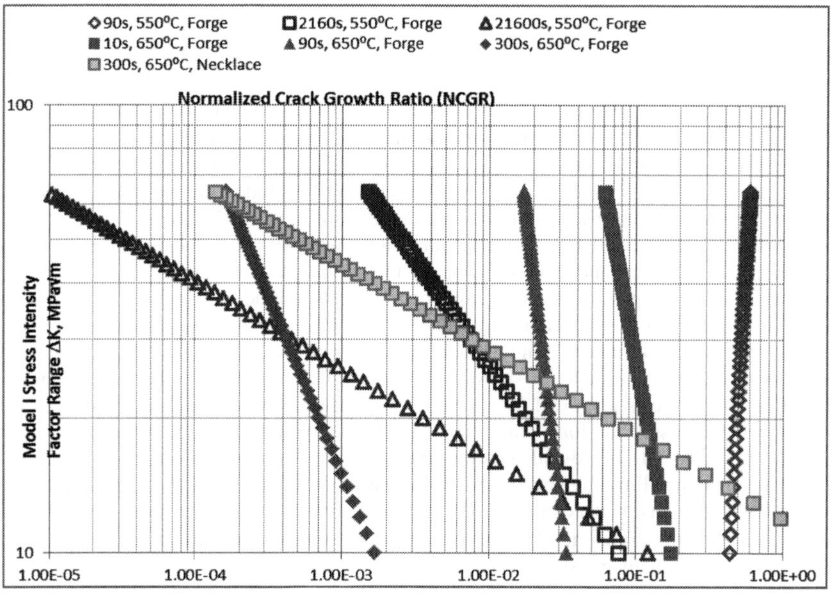

Figure 9: Dwell sensitivity map of Inconel 718 at high temperature

Inconel 718 shows accelerated crack growth at 650℃ at all dwell times, from 10s. The forged conditions exhibited higher crack growth resistance than the necklace structure. The 300s dwell at 650℃ exhibited similar trend $\Delta K = 11.923\ (NCGR)^{-0.189}$ to that of 21,600s dwell at 550℃, represented by a power equation where $\Delta K = 6.613\ (NCGR)^{-0.196}$,however, the dwell cycles were beneficial at the lowest range of long crack, ΔK, 10MPa√m, however, as stress intensity factor increased, the NCGR decreased causing dwell sensitivity, both cases four orders of magnitude. The dwell times of 10-90s at both the 550 and 650℃ caused only a factor of 4-5 times the dwell sensitivity between the ΔK range of 10 to 65MPa for 300s and 2160s dwell cycles there was an order of magnitude difference with continuous fatigue crack growth rates.

Crack Growth Damage Mechanisms (Long Cracks)

The damage mechanisms in long crack growth (LCG) are controlled by the dwell times. Relative amount of striations on the crack surface was dictated by the length of the dwell. Generally, supersalloys accumulate deformation by a stress assisted process which may produce oxidation and subsequent intergranular fracture. A shift in the damage mechanisms from transgranular to intergranular damage occurs at low frequency (0.5Hz) in Inconel 718. Intergranular cracking in Inconel 718 is associated with diffusion of oxygen along the grain boundaries, faster cycles do not provide the time needed to begin the oxidation and/or grain boundary damage, thus transgranular cracking takes place. This time is represented in term of characteristic time, and used to distinguish between the small scale yielding and the extensive creep. Wanhill [56] empirically related this with Poisson's ratio, Young's Modulus and other parameters in terms of $t_c = K_f\,(1-v^2)/[(n+1)EC^*]$. It was also proposed from the load-line compliance data that the Inconel 718 exhibits elastic deformation predominating during dwell crack growth at temperatures up to 650℃. However, plasticity is highly localized during the dwell fatigue crack growth. As proposed in LCF dwell, the DTDP includes the saturated, peak and relaxed stress, each contributing to the crack growth of long crack differently. Peak loads are correlated with dwell fatigue crack inhibition that may cause localized yielding and plasticity to build up and may develop residual compressive stress at the crack tip. Likewise, as the frequency increases, under continuous fatigue scenario, the crack growth rate no longer considered frequency dependent. Only at very high frequency, where cyclic plastic deformation helps build the temperature in a material, temperature sensitivity occurs. The material shows increased crack growth rates. The upper limit of this frequency rate is not available in the literature at room and high temperature. The damage mechanisms for long crack growth rates in Inconel 718 summarized above is shown in Figure 10. The dwell data plotted in Figure 8 shows various cycle types and mechanisms controlling the crack growth is a function of crack growth rates and range of mode I stress intensity factor in Fig. 10. It is clear that at steady and slow crack growth rates and ΔK the transgranular fracture occurs, however, as the crack growth rate increases the mechanisms shift to intergranular, oxidation and creep based mechanisms. Outside of mechanistic envelop shown in Fig. 10, the da/dN-ΔK exhibit a tertiary behavior which may be creep-fatigue dominated and likely mechanisms may be by LCF.

Figure 10: Dwell sensitivity damage mechanisms in long crack growth of Inconel 718 at high temperature

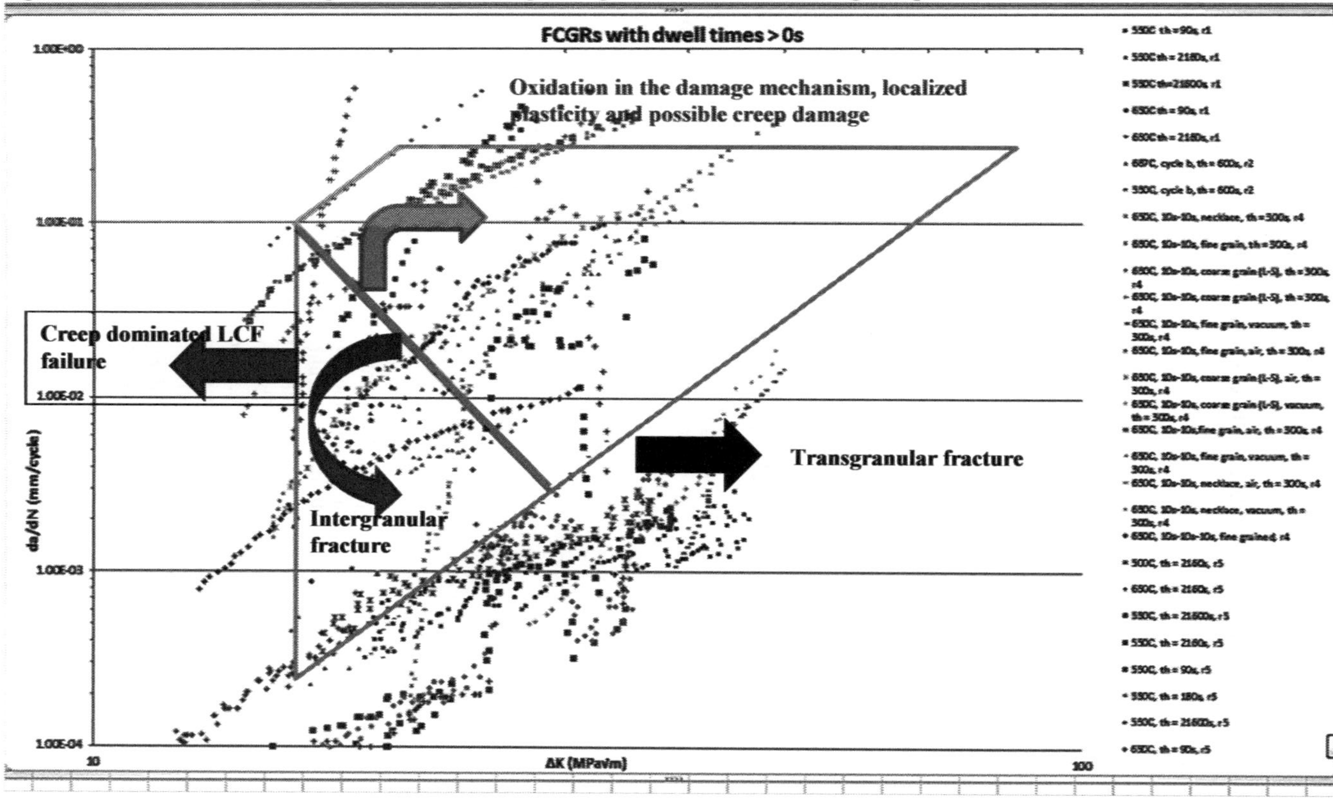

DWELL SENSITIVITY IN DISK LIFE ASSESSMENT

Dwell fatigue crack growth models comprising of 10 and 90s hold times were determined and applied to the finite element disk model life assessment discussed else-where [62-64]. The analysis assumed that the component had reached its' safe operational life or design criterion by safe life limits. At the end of component life a crack was assumed to have propagated at the critical locations of the disk namely: hub, bolt-hole and air-slot. The crack definitions were 0.79mmx0.79mm at the air-slot and hub, however, it was assumed 0.254mmx0.254mm in the bolt-hole. The Paris equation for the new material and for 10s and 90s dwell cycles were applied to the DARWIN program discussed in the accompanying paper [62 and 63-64] and presented in Table 2, below.

Table 2: Parameters of Paris equation for new material, and dwell cycles of 10 and 90s

"C" and "m" values for the Paris Equation

New		10 sec Dwell		90 sec Dwell	
C	m	C	m	C	m
1.44E-09	1.81	1.00E-11	4.6855	6.00E-10	2.9414

The sensitivity of dwell fatigue crack growth was obtained in terms of remaining life of the disk, based on a safe critical crack length definition. The disk shows sensitivity to dwell fatigue crack growth as below:

Table 3: Sensitivity of dwell fatigue crack growth in remaining life assessment

Flights to Failure

	New	10 sec Dwell	90 sec Dwell
Air-slot	1028	41	6
Hub	573	663	297
Bolt-hole	20001	5161	10013

From the table above it is quite clear that dwell fatigue crack growth lowers the component life considerably, in that, for the air-slot the life reductions were nearly 200 for 90s dwell, and 25 for the 10s dwell. For hub the life reductions were by one-half for 90s dwell, however, 10s dwell increased the life compared to the continuous 0-max-0 flight cycle by a factor of 1.15. The bolt-hole which had a smaller crack size, 0.25mmx0.25mm, the remaining life reduced from a factor of 4 for 90s to a factor of 2 for 10s dwell. This analysis found that there is a need to develop a general agreement from the non-destructive inspection (NDI) community particularly eddy current inspection about the crack size at the end of component life as well as other indications that characterize the material anomalies present on the surface and sub-surface of the material both at "short" and "long" sizes. These definitions would then determine the remaining life of the component. It is recommended that dwell sensitivity analyses be incorporated in risk assessment and life extension.

DISCUSSION

This paper demonstrates that dwell sensitivity occurs in strain control low cycle fatigue [5-50] as well as in long fatigue crack growth [51-60] scenarios. While a saturation in the number of cycles to failure with high strain range and dwell time noted in LCF, the dwell sensitivity was significant in fatigue crack growth as a result of high temperature and dwell time. Damage in Inconel 718 was found to be temperature dependent. In both the scenarios a normalized cycle ratio (NCR - LCF) or normalized crack growth ratio (NCGR - FCG) was able to describe the dwell sensitivity.

A safe component life is related with the LCF life which concentrates strains at localized regions. In these localized regions cracks form. Figure 4 shows the number of cracks and cyclic lives are directly related, in that, higher number of cracks forming on the specimen surface reduces the number of cycles to failure. Also, cracks and their numbers depend on the strain range applied, dwell time and temperature. Figure 4, caption, summarizes the strain range, number of cracks, and cycles to failure, it is evident that the highest number of cracks formed with the highest strains. The total strain range of 2.5% is very high and none of the components likely see in applications.

In rotating bending fatigue tests at room temperature, Inconel 718 exhibited reduced number of cracks as the cyclic life shifted from low to high cycle [65]. Within the low cycle regime, less than 10^5 cycles the number of crack origins increased from 5 or 6. However, as the cyclic life increased and failure in the high cycle, the number of crack origins decreased. Within the cycling life range of 10^6-10^8 cycles the crack origins were mainly via dominating fatigue crack. It may also be noted that these crack origins were on polished specimens and there was no interactions among the plasticity and surface oxidation due to exposure to high temperature. The damage mechanisms in room temperature fatigue of Inconel 718 are controlled by slip system in crystallographic sites that may be activated at multiple sites and/or grains [66]. These mechanisms were incorporated in the life assessment of turbine disk elsewhere [62-64]. At low stress, only the most easily operated slip systems can be introduced by localized plastic deformation during crack nucleation. Multiple crack origins exhibit lower life than the single crack fatigue behavior. At these levels, near ΔK_{th}, dwell sensitivity LCG was similar to no-dwell conditions.

The stress intensity factor (SIF) range at the end of "safe life" was determined from fracture toughness of the material and ranged from 6 - 20 MPa√m. For room temperature fatigue crack growth tests conducted in [66] found the SIF range to be 9-15 MPa√m. The crack grows in LCG regime beyond this point. A conservative estimate of the remaining life assessment reveals that initial crack size at the end of "safe life" may be too large to yield any potential life extension. The initial crack assumed was based on crack size at the end of LCF life considered to be 0.75mm long. The anomaly distribution and probability of detection of cracks by NDI methods are input parameters along with random crack definitions present at critical locations. These input parameters must be defined in order to effectively perform the remaining life of rotating components.

CONCLUDING REMARKS

The paper presented a summary of the dwell sensitivity behavior of high temperature materials. Dwell sensitivity was defined when the dwell fatigue resistance was lower than the no dwell or continuous fatigue resistance of materials at high temperature low cycle fatigue. The dwell sensitivity occurred at very low ranges of the strain and as the strain range increased, a saturation in the number of cycles to failure observed with further increase in dwell times. Some materials were found to be sensitive to dwell in peak tension loads, however, other materials exhibited compression dwell sensitivity.

Dwell sensitivity was described by a normalized cycle ratio (NCR) and NCR lower than 1 indicated dwell sensitivity in LCF. Material properties in monotonic fatigue and fatigue may characterize the material behavior under dwell. Strength ratio was found to be an indicator of dwell sensitivity behavior. Other parameters were developed and presented which measured dwell sensitivity. Dynamic stress-strain behavior in tension and compression direction was also found to predict the dwell sensitivity as well as new dwell time damage parameter proposed to measure dwell sensitivity.

Dwell sensitivity in fatigue crack growth was found more profound than in LCF. Data from the literature indicate that the new parameter proposed in this paper, normalized crack growth ratio (NCGR), characterized the dwell sensitivity. Inconel 718 exhibited dwell sensitivity at temperatures above 550C with long dwell times, however, at 650C smaller dwell times lowered the fatigue crack growth rates from an order-of-magnitude to four-orders-of magnitudes.

Damage mechanisms were elucidated for both LCF and LCG. A schematic map presented shows the mechanisms controlling the crack growth topography with TG, IG, mixed, oxidation, creep, cyclic plasticity build-up, and LCF. Combinations of crack growth rates and Mode I stress intensity factor range in each of these regimes be determined for Inconel 718 from the damage diagram and will be useful in design.

Remaining life assessment was performed for a model disk at the end of service life of the component. Various crack definitions were employed to show remaining life after the safe life of gas turbine engine disk. Dwell fatigue, LCG, reduces the remaining life of the component, where it may not be cost effective to extend the component life. However, there is a need for wide-spread collaboration and consensus development in the lifing community to defining the input parameters for the risk assessments.

ACKNOWLEDGEMENT

The author would like to acknowledge undergraduate students Bevin Duckett and Marcel Anthony, and research engineers, Jace Carter, and Heather Dwire with the Damage Tolerance and Probabilistic Life Prediction of Materials Center who perform the analyses.

References

[1] K. G. Wilkinson, Tech Air, 2 (1973) November Issue.

[2] C.A. Stuubington and S. Pearson, Eng. Fracture Mechanics, 10, 4, (1978) 723.

[3] J.H. Fielder and D. Birsch, The DC-10 Case, A study in applied ethics, technology, and society, State University of New York Press, Albany (1992).

[4] Department of Transportation, Federal Aviation Administration, Rule Vol. 63, No. 37, February 25, 1998 pp. 9405-9408, Airworthiness Directives; Pratt and Whitney JT8D Series Turbofan Engines.

[5] S.S.Manson Behavior of materials under conditions of thermal stress; Heat Transfer Symposium, University of Michigan, Engineering Research Institute, Michigan, (1953) pp. 9-75.

[6] L.F. Coffin, Jr. Trans. ASME, 76, (1954), pp. 931-950.

[7] Anon, Code Case N-47, (1976) ASME Boiler and pressure vessel code, Criteria for design of elevated temperature, Class 1 components in Stion III, Division 1, American Society of Mechanical Engineers.

[8] S.S. Manson, G.R. Halford, and M.H. Hirschberg, NASA Report TMX 67838 (1971).

[9] J. Lemaitre, J.L. Chaboche and Y Munakata, Method of metal characterization for creep and low cycle fatigue prediction in structures - Examples of Udimet 700, Proc. Symposium on Mechanical behaviour of materials, Kyoto, Japan, (1973), pp. 239-249.

[10] L.F. Coffin Jr., Fatigue at high temperature and interpretation, Proceedings of Institute of Mechanical Engineers, 9/74, (1974), p. 188.

[11] S. Majumdar, and P.S. Maiya, ASME/CSME Pressure Vessel and Piping Conference, PVP, PB 028, (1978)

[12] W.J. Ostergren, J. Testing and Eval. 4, (1976), pp. 327.

[13] N. Chrzanowski, I, Journal of Mechanical Science, 18, (1976), pp. 69-73.

[14] I.W. Goodall, and D.L. Thomas, Nuclear Electric Inc. Code R-5, (1990)

[15] D.R. Diercks, and D.T. Raskey, A statistical analysis of elevated temperature, strain controlled fatigue data of type SS 304 stainless steel, ASME Annual Winter Meeting, (1976), pp. 1-33.

[16] K. Sonoya, , I. Nonaka, and M. Kitagawa, Prediction of creep-fatigue lives of Cr-Mo steels with Diercks equation, ISIJ International, 31, (1991), pp. 1424-1430.

[17] Langer,

[18] S. Vaynman, M.E. Fine and D.A. Jeannotte TMS AIME Annual meeting, Denver, (1987), Ed. by Liaw, P. K. and Nicholas, T.

[19] Plumbridge, W. J. (1996-97) Private communications.

[20] R.H.Stentz, J.T. Berling, and J.B. Conway, AGARD CP 243 1978, paper no. 12

[21] E.G. Ellison, and A.J.F. Patterson, Proc. I. Mech. Engs. 190, (1976), 321

[22] T. Goswami, High Temperature Materials and Processes,14, (1995), 1, 1.

[23] Anon, NRIM Fatigue Data Sheet No. 28, National Research Institute for Metals, (1981) Tokyo, Japan.

[24] Anon, National Research Institute for Metals, Fatigue data sheet No. 62 (1989).

[25] Anon, National Research Institute for Metals, Fatigue data sheet No. 78 (1993).

[26] K. Yamaguchi, and K. Kanazawa, Met. Trans. 11A (1980), 2019.

[27] H.J. Shi, and G. Pluvinage, Int. J. Fatigue, 16, 9, (1994), 549
[28] J.B.Conway, R.H. Stentz, and J.T. Berling, Fatigue, tensile, and relaxation behavior of stainless steels, Technical Information Center, United States Atomic Energy Commission, (1975), pp. 33-81.
[29] G.R. Halford, L.R. Johnson, and J.A. Brown, High temperature LCF of Ni-201 and 304L S.S., NASA Conference Publication 2437, Advanced Earth-to-Orbit Propulsion Technology, Ed. R. J. Richmond and S. T. Wu, (1986), 172.
[30] J.M. Drapier, and M.H. Hirschberg, AGARD AR 130, (1979).
[31] W.J. Plumbridge, and M. Stanley, Int. J. Fatigue, 8, (1986), 209.
[32] K.D. Sheffler and G.S. Doble, Influence of creep damage on the low cycle thermal-mechanical fatigue behavior of two tantalum base alloys, (1972), NASA-CR-121001.
[33] V.T.A. Antunes and P. Hancock, AGARD CP 243, paper no. 5, (1978)
[34] E.G. Ellison, W.J. Plumbridge and M.S. Dean, Research Report 327, University of Bristol, Department of Mechanical Engineering, (1984).
[35] G. Asquith, and S.H. Springthall AGARD CP 243, (1978), paper no. 7.
[36] J.M. Hyzak and H.L. Bernstein AGARD CP 243, (1978), paper no. 11.
[37] C.S. Kortovitch and A.A. Sheinker, AGARD CP 243, (1978), paper no. 1.
[38] K.B. Rao, H.P. Meurer and H. Schuster, Mater. Sci. Engng. A104, (1988), 37
[39] G.R. Halford and A.J. Nachtigall, AGARD CP 243, (1978), paper no. 2.
[40] R.V. Miner, J. Gayda and M.G. Hebsur, ASTM STP 942, (1988), 371
[41] G.A.Swanson, I. Linask, D.M. Nissley, P.P. Norris, T.G. Meyer, and K.P. Walker, Life prediction and constitutive models for engine hot stion anisotropic materials program, NASA-CR-179594, (1987).
[42] T. Goswami, T. High Temperature Materials and Processes, 15, 1-2, (1995), 91-96.
[43] H.G. Edmund and D.J. White Observations of the effect of creep relaxation on high strain fatigue, J. Mech. Eng. Sci. 8, (1966) pp. 310-321.
[44] Y. Wada, Y. Kawakami, and K. Aoto, A statistical approach to fatigue life prediction for SUS 304, 316, and 321 austenitic stainless steels, Thermal stress, material deformation, and thermo-mechanical fatigue, ASME PVP 123, ed. H. Sehitoglu and S. Y. Zamrik, (1987), pp. 37-42.
[45] T. Goswami, Dwell sensitivity modeling – a new damage parameter, Materials and Design 25, 3, (2003), 191-197.
[46] T. Goswami and H. Hanninen, Dwell effects in high temperature fatigue behavior, paper I, Materials and Design, 22, 3, (2001), 199-215.
[47] T. Goswami and H. Hanninen, Dwell effects in high temperature fatigue damage mechanisms Paper II, Materials and Design, 22, 3, (2001), 217-236.
[48] T. Goswami, Dwell Sensitivity, Part I Behavior and Modeling, Mechanics of Materials 22, 2, (1996), 105-130.
[49] T. Goswami, Low cycle fatigue – dwell effects and damage mechanisms, International Journal of Fatigue, 21, 1, (1999), 55-76.
[50] T. Goswami and G.F. Harrison, Gas turbine disk lifing philosophies – a review, International Journal of Turbo and Jet Engines 12, 1, (1995), 59-77

[51] M Hörnqvist, T. Månsson and D. Gustafsson, High temperature fatigue crack growth in Alloy 718 – Effect of tensile hold times, Procedia Engineering, 10, (2011), 147-152.
[52] H. Andersson, C. Persson, and T. Hansson, Crack growth in IN718 at high temperature, International Journal of Fatigue, 23, 9, (2001) 817-827.
[53] J. Moverare and D. Gustafsson, Hold-time effect on the thermo-mechanical fatigue crack growth behaviour of Inconel 718. *Materials Science & Engineering A*, *528* (29-30), (2011), 8660-8670.
[54] J.P Pédron, and P. Pineau, The effect of microstructure and environment on the crack growth behavior of Inconel 718 alloy at 650°C under fatigue, creep and combined loading, Materials Science and Engineering, 56, 2, (1982), 143-156.
[55] D. Gustafsson, J.J. Moverare, S. Johansson, K. Simonsson, M. Hörnqvist, , T. Månsson, and S. Sjöström, Influence of high temperature hold times on the fatigue crack propagation in Inconel 718. International Journal of Fatigue, 33 (11), (2011), 1461-1469.
[56] R.J.H Wanhill, Significance of dwell cracking for IN718 turbine discs, International Journal of Fatigue, 24 (5), (2002), 545-555
[57] R. Tobler, Low temperature effects on the fracture behaviour of a nickel base superalloy. Cryogenics, 16 (11), (1976), 669-674.
[58] J.A. Lee, The effect of grain size upon the fatigue-crack propagation behavior of alloy 718 under hold-time cycling at elevated temperature, Engineering Fracture Mechanics, 25, 3, (1986), 305-314.
[59] M. Soboyejo, and C. Soboyejo, Micromechanisms of fatigue crack growth in a forged Inconel 718 nickel-based superalloy, Materials Science and Engineering: A, 270 (2), (1999), 308.
[60] O. Osinkolu, and G. Marchionni, Fatigue crack growth in polycrystalline IN 718 superalloy, Materials Science and Engineering: A, 356 (1-2), (2003), 425
[61] B.K. Duckett and T. Goswami, Dwell sensitivity in long fatigue crack growth and prediction model, internal memorandum, also submitted to International Journal of Fatigue (2012).
[62] J.A. Carter, M. Thomas, T. Goswami, and T. Fecke, Probabilistic risk assessment of a turbine disk, in the proceedings of International Symposium on Fatigue of Materials II: Advances and Emergences in Understanding, MST, October 2012, edited by *Dr. Ashraf Imam, Dr. T.S. Srivatsan and Dr. Raghavan Srinivasan,* published by The Minerals, Metals and Materials Society.
[63] J.A. Carter, T. Goswami and T. Fecke, Probabilistic risk assessment for life extension of turbine engine rotors, submitted to Materials and Metallurgical Transactions (2012).
[64] M.A. Thomas, J.A. Carter, L. Matson, T. Goswami, A. Whitney-Rawls and T. Fecke, Feature based discretization of a turbine disk for probabilistic risk assessment, submitted to ASME Journal of Gas Turbine and Power (2012).
[65] X. Ma, Z. Duan, H. Shi, R. Murai, and E. Yanagisawa, Fatigue and fracture behavior of nickel based superalloy Inconel 718 up to the very high cycle regime, Journal of Zhejiang University – Science A (Applied Physics and Engineering) 11, 10, (2010), 727-737.
[66] S. Suresh, Fatigue of Materials, Cambridge University Press, Cambridge, 1998).

Fatigue of Materials II Advances and Emergences in Understanding
Edited by: T.S. Srivatsan, M. Ashraf Imam, and R. Srinivasan
TMS (The Minerals, Metals & Materials Society), 2013

PROBABALISTIC RISK ASSESSMENT OF A TURBINE DISK

Jace A. Carter [a], Michael Thomas [a], Tarun Goswami [a], Ted Fecke [b]
a *Damage Tolerance and Probabilistic Lifing of Materials Center, Wright State University, Dayton, OH 45324, USA*
b *Aeronautical Systems Center/EN, Wright Patterson Air Force Base, OH 45433, USA*

Abstract

Current Federal Aviation Administration (FAA) rotor design certification practices risk assessment using a probabilistic framework focused on only the life-limiting defect location of a component. This method generates conservative approximations of the operational risk. The first section of this article covers a discretization method, which allows for a transition from this relative risk to an absolute risk where the component is discretized into regions called zones. General guidelines were established for the zone-refinement process based on the stress gradient topology in order to reach risk convergence. The second section covers a risk assessment method for predicting the total fatigue life due to fatigue induced damage. The total fatigue life incorporates a dual mechanism approach including the crack initiation life and propagation life while simultaneously determining the associated initial flaw sizes. A microstructure-based model was employed to address uncertainties in material response and relate crack initiation life with crack size, while propagation life was characterized large crack growth laws. The two proposed methods were applied to a representative Inconel 718 turbine disk. The zone-based method reduces the conservative approaches, while showing effects of feature-based inspection on the risk assessment. In the fatigue damage assessment, the predicted initial crack distribution was found to be the most sensitive probabilistic parameter and can be used to establish an enhanced inspection planning.

Keywords: Probabilistic; Initiation; Propagation; Life; Prediction; Extension; Turbine; Disk

Introduction

Turbine engine disks are designed for a finite life based on low-cycle-fatigue criterion and safe-life design approach. The safe-life approach considers components to be automatically replaced after 0.1% of the components have initiated a crack of 0.79mm (1/32 in) in length [1]. However, some premature material damage may occur due to handling, machining damage, fretting, or inherent defects on the surface as well as sub-surface causing cracks to propagate and become critical during the predicted "safe-life" [2]. Turbine disk failures resulting from anomalies, such as the catastrophic crash in Sioux City, IA in 1989, have led to the addition of damage tolerance approach in the life management process [3] mandating new non-destructive inspection methodologies to detect cracks. The damage tolerance approach assumes all components to contain inherent flaws, and inspection intervals are conservatively established based on fracture mechanics to ensure that flaws do not reach a critical size between inspection intervals. A safety factor is then applied to the predicted life to establish a safe inspection interval (SII).

Current design philosophies using safe-life and damage tolerance methods conservatively determine the operating life of disk components, retiring them regardless of a cause [4] either in terms of surface damage or presence of crack(s). As a result many of the components have useful life left in them. Therefore, the Retirement-for-Cause (RFC) method is being investigated by the U.S. Air Force to safely extend the operating life of components beyond traditional life limits to use more fully the useful life of disk components. The RFC method removes a component from service when a crack-like defect is detected during a Non-Destructive Inspection (NDI) [4]. Components can also be removed from service due to economic and/or risk limits. RFC can allow components to remain in-service longer and reduces the life-cycle costs. RFC utilizes damage tolerance concepts to establish inspection intervals to ensure that defects do not reach a critical size during service. However, the risk of failure will increase as components remain in-service beyond traditional LCF limits [5]. Therefore, the increase in risk must be accurately quantified before RFC can be safely implemented in the fleet life management of military aircraft engines.

Following the Sioux City accident, attempts were made to investigate the possibility of incorporating damage tolerance life method to account for inherent flaws for certification of new rotor designs [6-8]. A damage tolerance probabilistic design code, called DARWIN® (Design Assessment of Reliability With INspection) remains in development with Southwest Research Institute (SwRI). DARWIN® is a probabilistic fracture mechanics program which utilizes damage tolerance concepts to determine the Probability-of-Failure (POF) subject to low-frequency defects with and without inspection. It is an approved tool for new rotor design certification. However, current lifing methods outlined by the FAA and employed in DARWIN® account for defects prior to service and do not address component damage induced during service. Therefore, life management methods must address in-service damage before the RFC method can be safely implemented.

The probabilistic risk assessment has two paradigms; 1) relative risk which utilizes component as a whole and its failure risk algebraically summarizes the probability of failure at various locations in a disk such as bolt-hole, hub, air-slot etc, towards 2) an absolute risk, which looks into probability of failure at the most critical location. A method was proposed for accomplishing the absolute risk by discretizing a turbine disk for use in the probabilistic risk assessment. The discretization approach is currently employed in the risk assessment of embedded defects. However, discretization has not been applied to the risk assessment of surface defects in the component. Rather, current lifing methods utilize the life-limiting location of the life-limiting feature to assess component risk. The life-limiting location is based on the maximum stress location of the component, in which the max principal stress is extended through and applied to the entire surface area of the life-limiting feature [8]. Conversely, the discretization method utilizes the stress gradient topology to capture the effects of the stress gradient on the POF of the component. As the stress gradient of a feature increases, the discretization produces a lower risk, thus removing conservatism from the risk determination. This produces a more accurate, or absolute, risk value for each feature of the component. The methodology for establishing the discretization of the turbine disk, for use in RFC will be discussed.

A method for including low-cycle-fatigue damage in the probabilistic risk assessment of a high pressure turbine disk was developed to quantify the increase in risk as components remain in-service beyond the traditional limits. Total fatigue life is assumed to be the summation of crack initiation life and crack propagation life. A microstructure-based crack initiation model is utilized to develop a fatigue-induced anomaly distribution using variations in material properties. Crack propagation from an initial crack was determined using both short and long crack growth concepts. DARWIN® is used as the primary probabilistic module to determine the probability of failure with and without inspection. The risk assessment code was applied to Nickel based superalloy, Inconel 718 material, to assess the fatigue life of a representative turbine disk geometry.

2. Zone-Specific Risk Assessment

A feature-based discretization of method is necessary to move from the relative risk, of the life-limiting approach developed by the FAA, towards an absolute risk. Typically critical locations are considered to be the life-limiting location and are used in the risk assessment; however, metallurgical defects can be induced anywhere within the disk during the material melting process. Thus, the location of a defect can be considered to be a random variable to reduce a conservative life-limiting approach. The random location can be accounted for in the probabilistic analysis through a feature-based discretization method where the user subdivides the component's area into regions called zones. This allows every region to be probabilistically sampled during the analysis. The methodology assumes a low frequency of anomalies assuming one significant anomaly exists in the disk; thus, the probability of having multiple anomalies is assumed to be negligible.

2.1 Probabilistic Method

The probability of failure of a zone is modeled as independent events. The probability of failure of the disk can be obtained from the system reliability equation

$$P[disk] = P[fracture\ in\ any\ zone] = P[F_1 \cup F_2 \cup \ldots \cup F_N]$$
$$= 1 - \prod_{i=1}^{n}(1 - P[F_i]) \approx \sum_{i=1}^{n} P[F_i] \quad (1)$$

where F_i indicates the fracture of a defect origination in zone *i*, $P[F_i]$ indicates the probability of fracture of an initial defect located in zone *i*, N is the total number of zones, $\cup$ is the union logical operator, and the approximate equality occurs because $P[F_i]$ is small.

The probability of failure of a zone, i.e., $P[F_i]$ requires an anomaly be present and grow until it reaches the fracture toughness of the material. This can be represented as

$$P[F_i] = P[A_i]\ P[B_i|A_i] \quad (2)$$

where $P[A_i]$ is the probability of having an anomaly in zone *i* and $P[B_i|A_i]$ is the probability of failure given an anomaly in zone *i*. The probability of a defect occurring per surface area of material depends on the manufacturing procedure. Industry-developed exceedance curve defines the occurrence rate and size of defects [8]. The probability of having a defect in zone, i.e, $P[A_i]$, is a function of the ratio of the zone area to the total component area.

The probability of failure of a disk subject to a defect, $P[B_i|A_i]$, is computed using the probability fracture-mechanics life assessment in DARWIN®. It is assumed that the defects defined from the exceedance curve exhibits no crack formation time, i.e., crack initiation time is zero. The cycles to failure is estimated using DARWIN®'s crack propagation algorithm. The probability of failure is computed using either Monte Carlo sampling or importance sampling probabilistic methods [9-10].

2.2 Zone Refinement Method

The stress gradient topology is used in the zone-refinement process where defect locations are introduced in response to the stress gradient for each component feature. General guidelines for the discretization process has been established, which is only meant as a general starting point. Zones may require further refinement to attain convergence. It is desired that all defect locations in a feature exhibit equal risk when establishing the discretization [7]. Thus, each defect location should have similar risk contribution factors (RCF). In order to reduce computational time, zone refinement should be applied only to zones that significantly influence the overall disk risk. This is determined through the use of the RFC for a zone, which indicates the risk contribution of the zone relative to the risk of the disk, given as

$$RCF_i = \frac{P[F_i]}{\sum_{i=1}^{n} P[F_i]} \tag{3}$$

where N is the total number of zones. The risk contribution factors are defined such that the sum of each RCF will be of unity. Thus, a percent threshold can be assigned to determine if a zone should be refined. Through this approach, zones are refined until they have an RCF less than an RCF threshold.

As a general guideline for zone refinement, it was established that in the high stress gradient regions, where the stresses produce stress intensities above the stress intensity threshold, for every 1 ksi (6.895 MPa) in the transverse direction of the feature a defect location set was necessary. Meanwhile, in low stress gradient regions, where the stresses produce intensity factors below the stress intensity threshold, a defect location set is necessary only every 15 ksi (103.421 MPa) in the transverse direction of the feature. The defect location sets should contain corner crack type defect locations at the edges of the feature and any number of surface crack type defect locations on the surface between the edges. It was determined that for a variation of 1.5 ksi (10.342 MPa) thru the circular feature one surface defect was required for the defect location set. These guidelines were utilized in the establishment of the defect locations for both the circular and non-circular component features. For these features the line of symmetry was utilized to establish the transverse direction for the guidelines.

2.3 Numerical Example

A representative Inconel 718 turbine disk was used to demonstrate the zone refinement method. The three-dimensional model, which is characterized by eight major features as shown in Figure 1. The defect locations were introduced in response to the stress gradient topology of each feature to establish the feature-based discretization of the component as discussed in the previous section. Other inputs used in the probabilistic analysis are shown in Table 1. Finite element analysis was performed using Abaqus/CAE®. The material properties of Inconel 718 were obtained from the High Temp Metals Inc. [11] and Military Handbook, MIL-HDBK-5H [12]. Once the finite element model, material properties, and loading/boundary conditions were established, the stress analysis was performed. A convergence study was performed on the FEA to ensure convergence of the results.

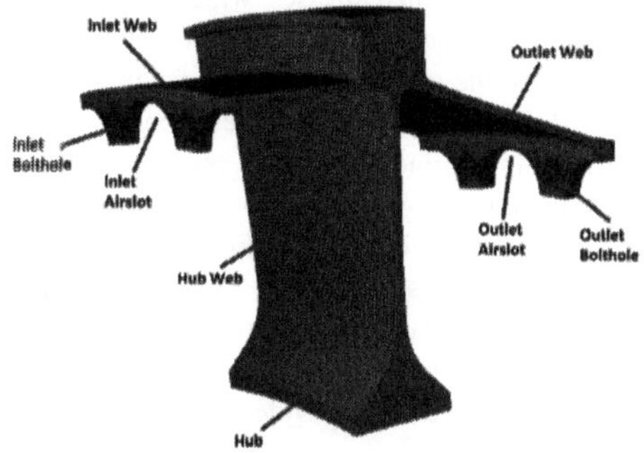

Fig. 1: Turbine disk model feature nomenclature.

Table 1. Input Values for the Probabilistic Analysis

Variable	Input Value
Fracture Mechanics Method	Flight_Life -Sigmoidal - No mean stress
Sigmoidal Crack Growth Model [13]	$\frac{da}{dN} = e^{-16.0397}\frac{\Delta K^{3.1911}}{\Delta K_{th}}\left(ln\frac{\Delta K}{\Delta K_{th}}\right)^{0.1}\left(ln\frac{\Delta K_c}{\Delta K}\right)^{-0.1424}$
Risk Computational Method	Monte Carlo Simulation @10,000 samples
Failure Toughness Factor	76.92 MPa$\sqrt{m}$ [13]
Stress Intensity Threshold	7.69 MPa$\sqrt{m}$ [13]
Stress Multiplying Factor	0
Defect Distribution	Manufacturing anomalies in circular features [8]
POD Curve	400 MV 50% Eddy Current [8]
Life Scatter	0
Inspection Time	4,000 & 8,000 cycles
Inspection Scatter	0

The introduction of the defect locations are both geometry and stress gradient dependent. The hub and hub web were found to have univariant stress gradients, which results in a single defect location set required through these features. These were the only two features which were found to have univariant stress gradients. Thus, dictating the development of an extensive method for establishing the feature-based discretization of the remaining features; to ensure that the defect location surface areas meet the assumption of univariant stress gradient at each defect location. The max principal stresses and the corresponding defect locations for the two life-limiting features have been provided in Figures 2 and 3.

The inlet air slots and bolt holes were found to have bivariant stress gradients. This produces the need for multiple defect location sets through the features, as shown in Figure 3 for the inlet air slot, to meet the assumption that the stress gradient of each defect location should be univariant. These features were also found to be symmetric about the transverse direction of the feature with respect to the max principal stresses, which can also be seen in the stress figure of Figure 2. The symmetry of the feature stress gradients allows for the establishment of fewer defect locations to fully define the total surface of the feature. This leads to an increase in the computational efficiency of the probabilistic risk assessment.

The symmetry of the max principal stresses and the stress gradients were also identified in the inlet web. As was found with the inlet air slots and bolt holes this decreases the number of defect locations needed to define the feature, thus increasing the computational efficiency of the model. The bivariant stress gradients of the inlet web lead to the necessity of multiple defect location sets to generate the discretization such that the surface area associated with each defect location will have a univariant stress gradient. The max principal stresses and corresponding defect locations of the inlet web have been provided in Figure 2.

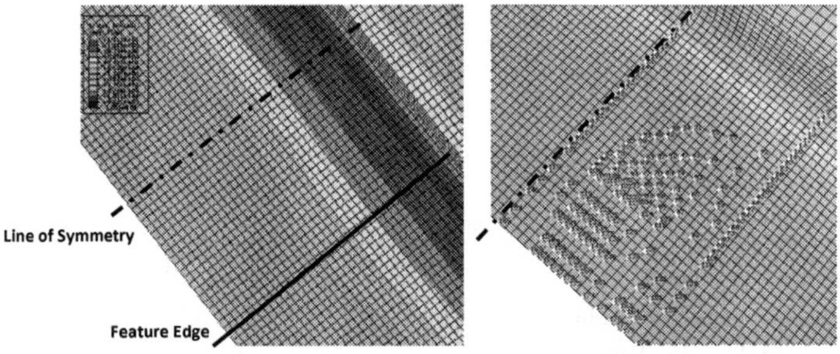

Fig. 2: Maximum principal stresses and the defect locations in inlet web.

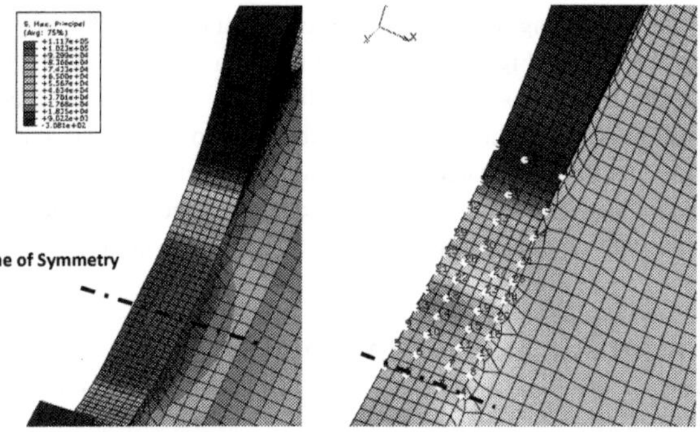

Fig. 3: Maximum principal stresses and the defect locations in inlet air slot.

A convergence study was performed on the number of defect locations within a given feature. However, there is more to the convergence study than just adding more defect locations. According to Millwater *et. all*, the identification of the defect locations requiring refinement is a key ingredient for the refinement [14]. For the probabilistic risk assessment, the defect locations should have nearly equal risk contribution factors (RCF). An initial run was performed for each feature and the feature POF was obtained. From the risk results of the probabilistic risk assessment a pie chart of the defect locations RCF's was created, i.e. the inlet air-slot RCF's shown in Figure 4. Defect locations which have RCF's greater than 5% are candidates for refinement. The probabilistic risk assessment of the inlet air-slot, containing 30 defect locations, has a large piece of the pie associated with one defect location, which is well above the 5% RCF limit. To reduce the RCF at this defect location, the area associated with the defect location was split into two and another defect location was added to the feature.

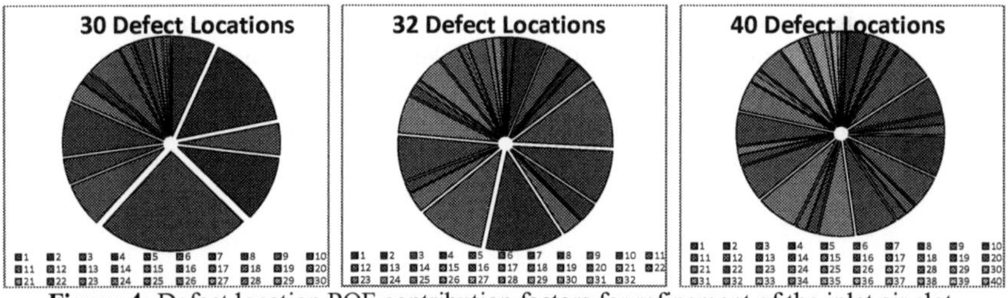

Figure 4: Defect location POF contribution factors for refinement of the inlet air slot.

This method of refinement was performed for all defect locations with risk contribution factors greater than the 5% RCF limit. The new surface areas were then determined and associated to the existing and new defect locations. Then, a new probabilistic risk assessment was performed. By investigating the effect of the refinement on the POF per flight cycle a check for convergence can be made. Figure 5 shows the POF versus flight cycles for the probabilistic risk assessment of the inlet air slot. From the figure, it can be seen that the POF has not converged to a solution between the 30 and 32 defect location risk assessments. Therefore, the defect location refinement was repeated, generating 40 defect locations. It can be seen that at this point the defect location RCF's are nearly equal with no exceedance of the RCF limit. Correspondingly, the POF demonstrates convergence to a POF solution, as can be seen in Figure 5.

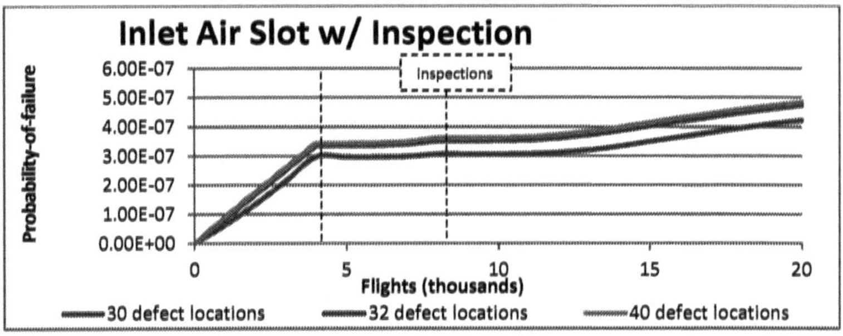

Fig. 5: Convergence Plot of POF versus Flights for inlet air slot with inspection.

This confirms that the convergence of the feature POF has been attained. The convergence procedure was repeated for each of the features that make up the representative turbine disk. Once the Convergence of the Monte Carlo simulation and the feature risk of the component have been obtained, the representative turbine disk POF can be established, in accordance with Equation 1. This will generate the overall risk for the component as seen in Figure 6. Table 2 provides the POF and # defect locations for each feature of the discretized turbine disk at 20,000 flight cycles. The turbine disk end life POF was found to be 1.94E-6. The probabilistic risk assessment was also performed utilizing the FAA life-limiting method. The life-limiting method resulted in an end life POF of 2.06E-6. The risk results of the two methods exhibit a 5.8% decrease in the end life POF, as a result of the discretization. The discretization method removes conservatism in the risk determination of the life-limiting method by accounting for the stress gradient topology.

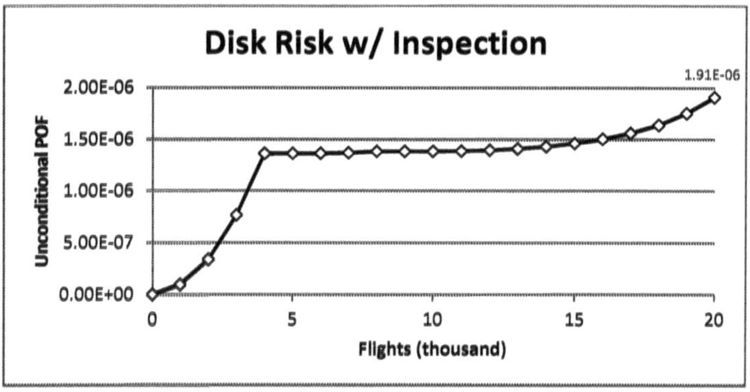

Figure 6: Plot of total disk risk with in-service inspections.

Table 2: Probabilistic risk assessment results for discretized turbine disk.

Feature	External Method	Defect Locations
Inlet Air Slot	4.84E-07	40
Hub	6.13E-07	5
Hub Web	5.42E-07	21
Inlet Web	2.42E-07	179
Outlet Air Slot	2.25E-08	39
Inlet Bolt Hole	3.25E-08	29
Outlet Web	4.71E-10	66
Outlet Bolt Hole	8.21E-09	17
Component Total	**1.94E-06**	**396**

3. Low-cycle Fatigue Damage

Although the zone-based method provides accurate risk estimation, fatigue damage must be addressed as components reach their LCF life. A method for including low-cycle-fatigue damage in the probabilistic risk assessment is used to quantify the increased risk as components remain in-service beyond the traditional limits. Total fatigue life is assumed to be the summation of crack initiation life and crack propagation life. A microstructure-based crack initiation model is utilized to develop a fatigue-induced anomaly distribution using variations in material properties. Crack propagation laws determine the life of the initial flaw to failure. DARWIN® is used as the primary probabilistic module to determine the probability of failure with and without inspection.

3.1 Fatigue Life Models

The total fatigue life (N_f) of representative turbine rotors was assumed to be comprised of two parts: (1) number of cycles to initiate a crack of a specific characteristic length and (2) number of cycles to propagate a crack to a critical length. The total fatigue life is represented as,

$$N_f = N_i + N_p \tag{4}$$

Where initiation life (N_i) and propagation life (N_p) both depend on the stress range, temperature, microstructure, and environment [15].

The fatigue-life variation in turbine engine materials have been investigated in several studies [15-17]. These studies indicate the importance of microstructure properties in the fatigue life. Tanaka and Mura [18] developed a crack initiation model that explicitly addressees microstructure properties. The model was developed by considering the dislocation-dipole mechanism along the slip band operating in a surface or subsurface grain, which ultimately leads to crack nucleation. Tanaka and Mura's model was recently extended by Chan *et al.* [19] to include crack initiation depth (a_i) and other relevant microstructure parameters.

$$(\Delta\sigma - 2\Delta\sigma_e)N_i^{\alpha} = \left[\frac{8M^2\mu^2}{\lambda\pi(1-\nu)}\right]^{1/2}\left(\frac{h}{D}\right)\left(\frac{a_i}{D}\right)^{1/2} \tag{5}$$

where $\Delta\sigma$ is the stress range, $\Delta\sigma_e$ is the endurance limit, μ is the shear modulus, λ is a universal constant (0.005), h is the slipband width, D is the grain size, and α is not a constant but depends on the degree of slip irreversibility and the stacking-fault energy, where $0 < \alpha \leq 1$. Upon crack initiation, the crack is assumed to grow according to large crack growth laws.

3.2 Probabilistic Fatigue Framework

A probabilistic framework has been developed utilizing the fatigue crack growth models and DARWIN® to determine the probability of failure due to low-cycle-fatigue damage as shown in the schematic (Figure 7). A probabilistic code was developed in Matlab to characterize fatigue damage in the form of an anomaly distribution using the initiation equation (Eq 2). The fatigue anomaly distribution is input into DARWIN® to determine the probability of failure with and without inspection. Random variables were assumed to be statistically uncorrelated, thus only certain variables were chosen to have a distribution. The fatigue damage program assumed that a single crack located in the highest stress region will dominate component failure since multiple micro cracks will either coalesce into a single crack or one crack will dominate over the others. Thus, only a single crack is assumed to exist for each of the major regions of the component. A more thorough discussion about the method can be found in previous work [20].

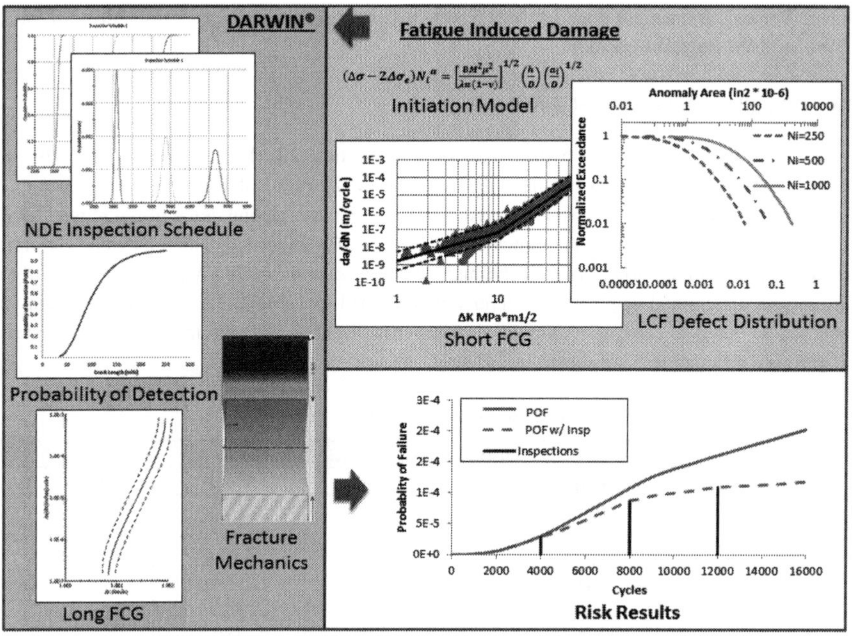

Figure 7: Probabilistic framework

A Monte Carlo simulation was used to determine the initiated crack size distribution using Equation 5. In this study, lognormal distributions were used for unknown variables since fatigue life has been shown to be lognormal [21]. The mean and standard deviation of the random material properties were obtained from experimental data from the literature. Since the variables were lognormal random variables, the logarithmic of the variables were Gaussian random variables. The mean and standard deviation of the logarithmic distributions were represented by equations 8 and 9, respectively.

$$\mu_{iL} = 2\log(\mu_i) - \frac{1}{2}\log({\mu_i}^2 + {\sigma_i}^2) \tag{6}$$

$$\sigma_{iL} = [-2\log(\mu_i) + \log({\mu_i}^2 + {\sigma_i}^2)]^{\frac{1}{2}} \tag{7}$$

where μ_i and σ_i are the mean and standard deviation of the random variable, respectively. Once the initiated crack distribution is determined, it was submitted to DARWIN® in the form of an anomaly distribution. Since the crack size was a random variable in Equation 2, the cycles to initiation (N_i) held constant and assumed to be correspond to the lower bound (-3σ) on the LCF curve. The risk assessment code in DARWIN® was used to determine the probability of failure of the initiated crack distribution.

3.4 Application to a Nickel Turbine Rotor

The Inconel 718 turbine disk discussed in Section 2.3 (Figure 1) was run using the purposed method for the two life-limiting locations of the disk, inlet air slot and inlet web. A surface crack located at the air slot was assumed to be the dominating crack in the region, which corresponds to the location of the highest stress of 965 MPa (Figure 2). The inlet web is assumed to have a dominating corner crack at the highest stress location of 700 Mpa. The probability of failure was determined using various inputs as described in Table 1. The loading spectrum was assumed to have a zero to maximum loading per flight cycle and microstructure material properties for Equation 5 are shown in Table 3.

Table 3: Inconel 718 material Properties

Initiation Model [22]				
Variable	**Description**	**Units**	**Mean**	**Stad. Deviation**
D	Grain Size	μm	6.7	2.68
M	Taylor Factor	-	2	-
λ	Universal Constant	-	0.005	-
μ	Shear Modulus	MPa	8.25E+04	-
ν	Poisson Ratio	-	0.3	-
$\Delta\sigma_e$	Endurance Limit	MPa	480	-
α	Life Exponent	-	0.204	-
h	Slipband Width	μm	1.0E-3	-
Propagation Model [13]				
B	Sigmoidal Constant	-	-16.0397	-
P	Sigmoidal Constant	-	3.1911	-
Q	Sigmoidal Constant	-	0.1	-
D	Sigmoidal Constant	-	-0.1424	-
K_{TH}	Stress Intensity Threshold	MPa $\sqrt{m}$	7.69	-
Kc	Fracture Toughness	MPa $\sqrt{m}$	76.92	-
R	Stress Ratio	-	0	-

The fatigue damage program created and analyzed the fatigue anomaly distribution for the two locations using the initiation life (Ni) of 4,000 and 6,000 cycles for the air slot and web, respectively. A Monte Carlo simulation of 10,000 samples was used to determine the initiated crack size distributions modeling grain size as lognormal random variable where other material properties are assumed to be constant. The initiated fatigue crack distributions for the two locations are given in Figure 8.

The POF analysis was performed using the two defect distributions with the sigmoidal large crack growth equation. Figure 8 shows that the two anomaly distributions have a large impact on POF assessment. The accuracy of risk assessment is directly related to the accuracy of the input distributions. However, not all input distributions are equally important. This study showed that the anomaly distribution is the most sensitive parameter. Figure 9 demonstrates how the initiated crack sizes increase exponentially with increasing initiation life, which is consistent with Equation 5 showing that the mean initial crack size is proportional to $N_i^{2\alpha}$. In comparison, the impact of other input distributions was relatively small. However, these conclusions are not general and it is recommended that other factors may be included in component risk assessment. Accurate assessment of component risk must consider not only material defects, but also the uncertainty associated with inspection time, probability of detection by nondestructive evaluation, and mission histories. The purposed probabilistic method coupled with DARWIN® has the capability to address uncertainties and is useful in determining the risk associated with component life extension. Furthermore, the predicted crack size distribution can be used to establish an enhanced inspection schedule to ensure an adequate risk level is maintained.

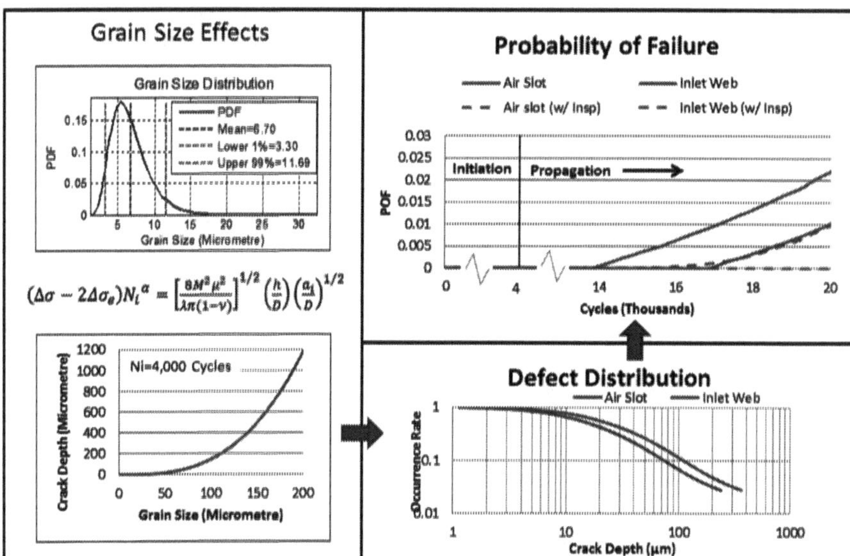

Figure 8: Risk results for the two selected features also showing how increasing grain size has an adverse effect on crack initiation

4. Conclusions

Turbine rotor disks have been traditionally managed by the combination of safe-life and damage tolerance methods. These methods are conservative and yield high life-cycle cost, thus a probabilistic RFC method is being investigated by the U.S. Air force to safely manage components beyond their traditional LCF limit. The discretization method provides an accurate way to assess the component risk due to rare defects by allowing multiple areas of the disk to be sampled instead of addressing only the life-limiting location. For the test case in this study, the discretization method provided a 5.8% decrease in the end of life POF compared to the FAA life-limiting method. The discretization method was close to the life-limiting method; however, there are many possible factors which may have led to this result. For example, the web features utilized the same default anomaly distribution as the circular-hole features. The difference was that the occurrence rate was established based on experience empirically. A change to the web anomaly distribution or frequency reduction factor could result in a change in the similarity of risk values. Therefore, a conclusion cannot be easily drawn rather it could be a coincidence of the approximation of some of the analysis parameters of the probabilistic risk assessment. However, the risks of the features are lower for the discretization method than that of the life-limiting approach. This occurred because the discretization method accounts for the stress gradient topology which removes some level of conservatism from the risk assessment in an effort to move towards an absolute risk value. In addition, the method provides the user with a robust risk assessment. For example, effects of feature-based inspections on the risk can be obtained by using different parameters (i.e. POD curves or defect occurrence rates) for the various discretized features.

Fatigue damage is another aspect to the RFC method that must be addressed before components can be safely managed beyond their traditional LCF life. A probabilistic fatigue life model developed to quantify the risk of failure as components reach their LCF life was applicable in fleet life management. A microstructure-based initiation model relates crack initiation life to crack depth by using grain-size material parameters. A fatigue damage risk assessment code that integrated the fatigue life model into the probabilistic code DARWIN® was able to predict the probability of failure with and without inspection. The initiation model was used to determine the initial crack size distribution used in the probabilistic fracture mechanics. The methodology was applied to Inconel 718 turbine disk to determine the probability of failure due to fatigue induced damage. Results showed that the crack size at which the crack is assumed to be initiated is the most sensitive parameter. The advantage of the fatigue damage program is its usefulness in estimating the expected initial crack size for the POF analysis, which can further be used to establish an enhanced inspection planning.

5.0 Acknowledgements

This work was supported by Universal Technology Corporation via sub-contract agreement12-S590-0016-11-C3. Ashley Whitney-Rawls from Wright Patterson Air Force Base provided continuous support of this work.

References

[1] R.M. Kearsey, A.K. Koul, J.C. Beddoes and C. Cooper: The minerals, Metals & *Materials Society*, 2000, pp. 117-126.

[2] S. Suresh: *Fatigue of Materials*, 2nd ed., Cambridge University Press, Cambridge, UK., 1998.

[3] J. Kappas: DSTO-TR-1306, 2002, Aeronautical and Maritime Research Laboratory, Australia.

[4] J-P. Immarigeon, A.K. Koul, and P.C. Patnaik: NATO-RTO-MP-079(11) Lecture Series, October 2001.

[5] S. Vukelich: NATO-RTO-MP-079(11) Lecture Series, October 2001.

[6] H. Millwater, M. Enright, and S. Fitch: *Journal of Engineering for Gas Turbine and Power*, 2007, vol. 129, issue 3, pp. 827-835.

[7] Federal Aviation Administration: *Advisory Circular*, 2001, Section 33.14

[8] Federal Aviation Administration: *Advisory Circular*, 2009, Section 33.70-2

[9] Wu, Y.-T., Enright, M. P., and Millwater, H. R., 2002, "Probabilistic Methods for Design Assessment of Reliability With INspection," AIAA J., 40_5_, pp. 937–946.

[10] Huyse, L. J., and Enright, M. P., 2003, "Efficient Statistical Analysis of Failure Risk in Engine Rotor Disks Using Importance Sampling Techniques," *44th AIAA/ASME/ASCE/AHS/ASC Structures, Structural Dynamics, and Materials Conference*, Norfolk, VA.

[11] High Temp Materials, "Inconel 718 Technical Data", High Temp Materials, September 2011, <http://www.hightempmetals.com/techdata/hitempInconel718data.php>

[12] Metallic Materials and Elements for Aerospace Vehicle Structures", MIL-HDBK-5H, 1998

[13] Whitney-Rawls, Ashley, "Impact of Induced Defect on Rotor Life Assessment", Wright State University, 2010.

[14] Millwater, H., Enright, M., Fitch, S., "Convergent Zone Refinement Method for Risk Assessment of Gas Turbine Disks Subject to Low-Frequency Metallurgical Defects", Journal of Engineering for Gas Turbine and Power, July 2007, Vol. 129.

[15] K.S. Chan and M.P. Enright: *Metallurgical and Materials Transactions A*, 2005, vol. 36, Issue 10, pp.2621-2631.

[16] S.D. Antolovich and N. Jayaraman: *Fatigue: Environment and Temperature Effects*, 1983, Plenum Press, NY, pp. 119-144.

[17] K. Kobayashi, K. Yamaguchi, M. Hayakawa, and M. Kimura: *Materials Letters*, 2005, vol. 59, pp. 383-386.

[18] K. Tanaka and T. Mura: *Journal of Applied Mechanics-Transactions of the ASME*, 1981, vol. 48, pp. 97-103.

[19] K.S. Chan: *Metallurgical and Materials Transactions A*, 2003, vol. 34, pp. 43-58.

[20] J. Carter, T. Goswami, and T. Fecke: Submitted to the *Journal of Metallurgical and Materials Transactions A*. 2012.
[21] C. Annis: *Journal of ASTM International*, 2004, vol.1, No. 2
[22] M.P. Enright and K.S. Chan: Journal of ASTM International, 2004, vol. 1, No.8.

Fatigue of Materials II Advances and Emergences in Understanding
Edited by: T.S. Srivatsan, M. Ashraf Imam, and R. Srinivasan
TMS (The Minerals, Metals & Materials Society), 2013

Simulating Fatigue Cracks in Healthy Beam Models for Improved Identification

Phillip E. Cooley[1], Joseph C. Slater[2], Oleg V. Shiryayev[3]

[1]Graduate Research Assistant; Wright State University, Dayton, OH, USA
[2]Associate Dean and Professor; Wright State University, Dayton, OH, USA
[3]Assistant Professor; Petroleum Institute, Abu Dhabi, United Arab Emirates

Keywords: Fatigue Crack, Nonlinear, Beam Models, Identification

Abstract

There is need for an automated Structural Health Monitoring (SHM) system capable of fatigue crack detection in bladed disks as current methods are slow, costly and imperfect. Prerequisite for such a system is a fast method for producing the necessary data libraries. In an effort to develop such a method for simulating nonlinear structural response, fatigue cracks in beams have been modeled by modifying the inputs to the structure rather than the structure model itself to produce a closed-form solution for the total response. Although the time savings are enormous, and the method has proven capable of correctly identifying fatigue cracks over an effective region in data generated by a more traditional bilinear model, additional refinement is needed. The method for calibrating the signature profiles used to identify fatigue cracks between methods is revisited and successfully improved. Existing signature profiles for the bilinear model are reproduced with higher resolution, and new features are observed. Different boundary conditions are evaluated with the new method and compared to published results. Although similar, there are still discrepancies that remain and will need to be investigated. Overall, the proposed method for modeling and identifying fatigue cracks in beams has been improved, but will require validation against physical experiments before being used on more complicated structures such as bladed disks.

Introduction

Fatigue cracks are well known to cause failure of components in turbomachinery, often with disastrous results [1, 2]. To avoid catastrophic failure due to fatigue, much effort has been directed toward improving the safety and reliability of the components used in aircraft turbomachinery [3]. In spite of these efforts, fatigue cracks remain difficult to identify even with rigorous inspection, and the cost of traditional maintenance inspections for turbomachinery components is high. To resolve these issues, the development of an automated SHM system would greatly decrease maintenance costs for turbomachinery while simultaneously increasing the likelihood of preventing failure.

Despite the difficulty in spotting fatigue damage, there are unique characteristics that can identify the presence of a fatigue crack. As a structure is exposed to vibration, nonlinear elastic behavior results from the crack opening and closing. Although the nonlinearity is highly localized around the crack, its behavior is well documented in beams [4]. Methods have been developed and verified experimentally that exploit this phenomena through the changes in Randomdec signature statistics resulting from the onset of crack nonlinearity. [5, 6, 7]. Other methods and nonlinear features have been investigated as well [8, 9, 10, 11]. To facilitate the development of an automated SHM system capable of identifying fatigue crack location and severity in bladed disk assemblies, faster methods for obtaining the response must be developed to generate sufficiently large data libraries. Previous efforts have investigated such a method for fatigue cracks in beams to create a framework for modeling and analysis in bladed disks. The focus of this research is to continue investigating the proposed method and to suggest changes that may improve its effectiveness. A summary of both an earlier method used for comparison and the proposed method itself is provided in the background section. An overview of the items investigated are provided in the next section, followed by the results and conclusion.

Background and Previous Work

The response of a linear system excited at a single driving frequency is characterized by oscillation at the same frequency. However, if a fatigue crack is present, the system is no longer linear and harmonics will be observed in the response. To more readily identify the presence of harmonics, the response can be analyzed in the frequency domain where it is easier to isolate magnitude and phase. The auto spectrum density (ASD) is a particularly effective tool for analyzing frequencies present in a signal, as shown in Figure (1) where signals from a healthy and cracked beam are contrasted.

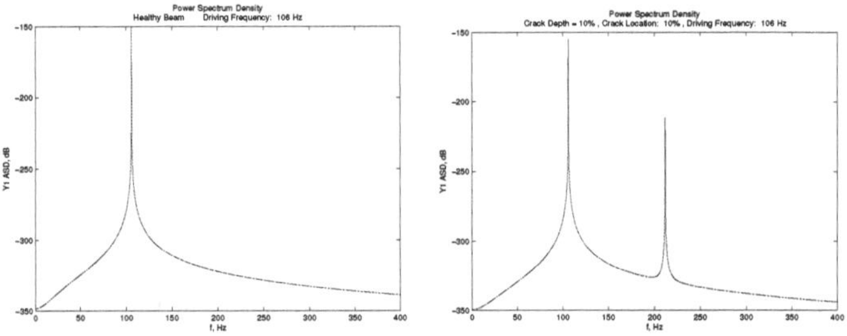

(a) Healthy Beam (b) Cracked Beam

Figure 1: The presence of a crack introduces additional frequency content to the response ASD as shown.

As a beam vibrates, it dynamically deforms to assume an operational deflection shape (ODS) dependent on the frequency of excitation. Such deformation induces various oscillating degrees of curvature along the length of the beam. At the location of the crack, curvature is of critical importance since it determines whether the crack is open or closed. For a crack on the top surface of a beam, negative curvature will open the crack introducing a point bending moment at the crack tip that produces the harmonics observed in the response ASD. Conversely, positive curvature will cause the crack to close, and the structure will behave as a healthy beam. Because of the bilinear nature of a beam with a breathing fatigue crack, one method solves two separate models and uses the curvature at the crack site to determine when to switch between them [12, 13]. This approach uses a torsional spring to geometrically represent the fatigue crack when it is open. An alternative closed-form approach was developed and compared to the torsional spring method in previous work[14].

Torsional Spring Approach

The closed-crack model is essentially a healthy beam, whereas the open-crack model includes an additional torsional spring that accounts for the increased compliance at the crack location. Examples of both closed and open cases are shown in Figures 2(a) and 2(b) respectively for the spring hinge-free configuration. The result is a pair of eigenproblems, each with a different set of mode shapes and natural frequencies. By monitoring the beam curvature at the crack location for each time step, the status of the crack can be determined and the appropriate model selected. The critical limitation of this method is that the mesh around the crack location must be very fine to ensure the correct model is used during curvature transition, and this results in extraordinarily large amounts of time needed to run the simulation. Since libraries of data must be generated, a faster method is imperative.

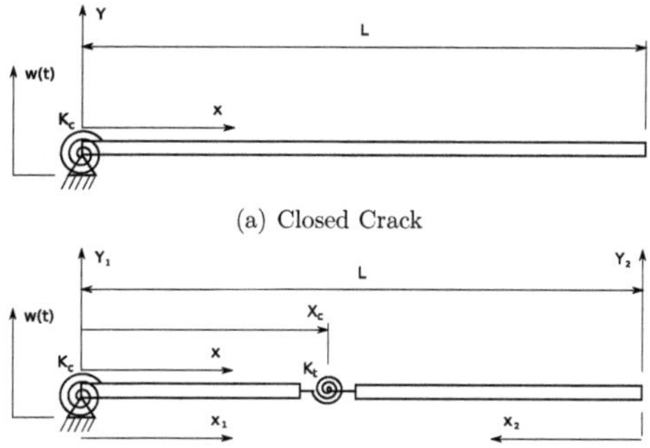

(a) Closed Crack

(b) Open Crack

Figure 2: Models of the spring hinge-free beam configuration for both closed and open crack states.

When the fatigue crack is closed, the system is linear and a continuous beam can be used to determine the mode shapes. When the fatigue crack is open, two continuous sections are used to solve for the mode shapes. However, since continuity conditions are introduced at the crack location for the open case, the resulting combined mode shapes for the open case are also continuous along the beam. As a result, the general derivation is the same but with the mode shapes and natural frequencies used at each time step determined by the beam curvature at the crack location. It has previously been shown[14] that the reduced governing equation is given by

$$\sum_n \left[\left(\ddot{q}_n(t) + 2\zeta_n\omega_n\dot{q}_n(t) + \omega_n^2 q_n(t) \right) Y_n(x) \right] = f_\ell(t) + \frac{F_{ext}(x,t)}{\rho A} \tag{1}$$

where $F_{ext}(x,t)$ is any external excitation, $f_\ell(t)$ represents forces induced by rigid body motion, and the displacement relative to the neutral surface in the beam is given by $\sum_n Y_n(x)q_n(t)$. Both rigid body motion terms and external excitation terms are included to permit ready use whether using a shaker or repeated loading at a point along the beam. Mass normalizing the mode shapes and taking advantage of orthogonality yields the modal equations in Equation (2).

$$\ddot{q}_n(t) + 2\zeta_n\omega_n\dot{q}_n(t) + \omega_n^4 q_n(t) = f_\ell(t)\int_0^L \rho A Y_n(x)dx + \int_0^L F_{ext}(x,t)Y_n(x)dx \tag{2}$$

where ζ_n and ω_n are the corresponding modal damping ratios and natural frequencies, and the right side of the equation represents modal forces resulting from rigid body motion and external excitations.

For the spring hinge-free configuration, one end of a beam is attached to a shaker. Since the bolts used to attach the beam are not fully rigid, the boundary condition for the attached end is considered a spring hinge rather than clamped. The shaker provides the input given in Equation (3), and since there is no other external excitation, $F_{ext} = 0$.

$$y_{rb}(t) = A\sin(\Omega_{dr}t) \tag{3}$$

To construct the eigenvalue problem for the spring hinge-free beam, the boundary conditions are considered first for the case of a closed crack in Figure 2(a)

$$\text{at} \quad x = 0: \qquad Y_n(0) = 0,\ EIY_n''(0) = K_cY_n'(0) \tag{4}$$

$$\text{at} \quad x = L: \qquad Y_n''(L) = 0,\ Y_n'''(L) = 0 \tag{5}$$

where K_c is the torsional stiffness of the spring hinge and primes represent spatial derivatives with respect to x. From the boundary conditions, the natural frequencies are obtained numerically. The closed-crack mode shapes have previously been obtained[14] and are given in Equation (6).

Closed-Crack Mode Shapes:

$$\begin{aligned} Y_n(x) \ = \ & c_n\Bigg[-(1+\sigma_n)\cos\beta_n x + \left(\frac{2\beta_n EI}{K_c} + 1 + \left(\frac{2\beta_n EI}{K_c} - 1\right)\sigma_n\right)\sin\beta_n x \\ & + e^{-\beta_n x} + \sigma_n e^{\beta_n x}\Bigg] \end{aligned} \tag{6}$$

where c_n is used to mass normalize the mode shapes, and coefficients σ_n are defined as

$$\sigma_n = \frac{\left(\frac{2\beta_n EI}{K_c}\right)(\sin\beta_n L + \cos\beta_n L) + 2\sin\beta_n L}{\left(\frac{2\beta_n EI}{K_c}\right)(-\sin\beta_n L - \cos\beta_n L) + 2\cos\beta_n L + 2e^{\beta_n L}} \tag{7}$$

When the crack is open, the eigenvalue problem is formed in order to calculate the mode shapes and natural frequencies by considering boundary conditions and continuity conditions at the location of the crack. The spring hinge is at the left end ($x_1 = 0$) while the right end ($x_2 = 0$) is free. The mode shapes in Equations (14) and (15) have been derived in previous work[14].

Open-Crack Boundary Conditions:

$$\text{at} \quad x_1 = 0: \qquad Y_1(0) = 0, \; \frac{d^2}{dx_1^2}[Y_1(0)] = \frac{K_c}{EI}\frac{d}{dx_1}[Y_1(0)] \tag{8}$$

$$\text{at} \quad x_2 = 0: \qquad \frac{d^2}{dx_2^2}[Y_2(0)] = 0, \; \frac{d^3}{dx_2^3}[Y_2(0)] = 0 \tag{9}$$

Open-Crack Continuity Conditions:

$$Y_1(x_c) = Y_2(L - x_c) \tag{10}$$

$$\frac{d^3}{dx_1^3}[Y_1(x_c)] = -\frac{d^3}{dx_2^3}[Y_2(L - x_c)] \tag{11}$$

$$\frac{d^2}{dx_1^2}[Y_1(x_c)] = -\frac{K_T}{EI}\left(\frac{d}{dx_1}[Y_1(x_c)] + \frac{d}{dx_2}[Y_2(L - x_c)]\right) \tag{12}$$

$$\frac{d^2}{dx_2^2}[Y_2(L - x_c)] = -\frac{K_T}{EI}\left(\frac{d}{dx_1}[Y_1(x_c)] + \frac{d}{dx_2}[Y_2(L - x_c)]\right) \tag{13}$$

Open-Crack Mode Shapes:

$$Y_{1n} = \delta_n\Big[- \mathcal{A}_n \cos\beta_n x + \left(2 - \mathcal{A}_n\left(1 - \frac{2\beta_n EI}{K_c}\right)\right)\sin\beta_n x + e^{-\beta_n x} + (\mathcal{A}_n - 1)\, e^{\beta_n x}\Big], \quad 0 < x < x_c \tag{14}$$

$$Y_{2n} = \delta_n \alpha_n\Big[- \mathcal{B}_n \cos\beta_n (L - x) - (\mathcal{B}_n + 2)\sin\beta_n(L - x) + e^{-\beta_n(L-x)} - (\mathcal{B}_n + 1)e^{\beta_n(L-x)}\Big], \quad x_c \leqslant x < L \tag{15}$$

where coefficients α_n are expressed as

$$\alpha_n = \frac{-\mathcal{A}_n \cos\beta_n x_c + \left(2 - \mathcal{A}_n\left(1 - \frac{2\beta_n EI}{K_c}\right)\right)\sin\beta_n X_c + e^{-\beta_n x_c} + (\mathcal{A}_n - 1)\, e^{\beta_n x_c}}{-\mathcal{B}_n \cos\beta_n(L - x_c) - (\mathcal{B}_n + 2)\sin\beta_n(L - x_c) + e^{-\beta_n(L-x_c)} - (\mathcal{B}_n + 1)e^{\beta_n(L-x_c)}} \tag{16}$$

Coefficients $\mathcal{A}_n$ and $\mathcal{B}_n$ are expressed as shown in the appendix, and δ_n are constants used to mass normalize the mode shapes.

Proposed Closed-Form Method

The proposed method summarized in this section was developed specifically to reduce the computation time for generating the cracked beam response to permit faster crack identification. By directly modeling the forces resulting from the presence of the fatigue crack, a single model can be used in place of the costly dual-model approach which eliminates the need for heavy mesh refinement in the area of the crack and increasingly smaller time steps as the curvature transitions. For repeated sinusoidal loading, the crack can be directly modeled as an applied bending moment that is only active when the curvature causes the crack to open. The Fourier series expansion of the half-wave rectified sinusoid shown in Figure (3) and Equation (17) is selected to represent the temporal component of the nonlinear self-excitation since a fatigue crack is essentially inactive when closed.

$$F_c(t) = \frac{1}{\pi} + \frac{1}{2}\sin(\Omega_d t) + \sum_{k=2}^{\infty} \frac{2}{\pi}\left(\frac{1+\cos(k\pi)}{1-k^2}\right)\cos(k\Omega_d t) \tag{17}$$

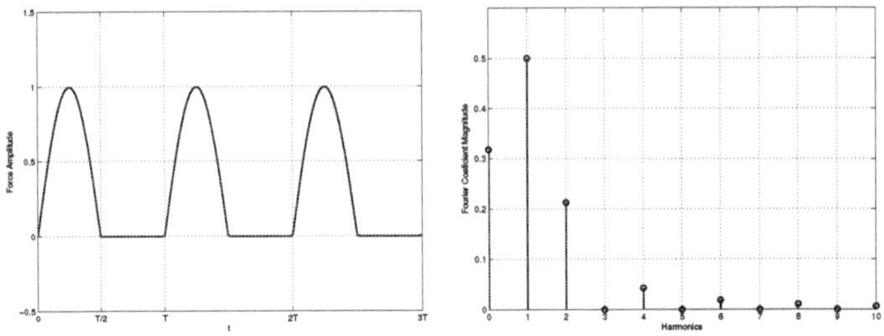

(a) Time Domain (b) Fourier Coefficients

Figure 3: Model of forces from crack surface interaction.

Observing the Fourier coefficients in Figure 3(b), the second harmonic is identified as the most prominent after the fundamental. Since the nonlinear features are relatively small, the driving frequency is set to half of any natural frequency so that the second harmonic coincides with the modal resonant peak. By defining the driving frequençy in this way, the nonlinear response is increased without jeopardizing the integrity of the structure. Also note in the stem plot of Figure (3) that although the harmonic at double the driving frequency is significant, there is no harmonic observed at half the driving frequency. This indicates that despite the prediction of a subharmonic resonance by the bilinear stiffness model [15], the nonlinearity introduced by a breathing fatigue crack will only produce superharmonic resonance. Although not anticipated, this behavior was observed in previous work by Meier et al.

Consider once again a beam with generic boundary conditions and inputs. A crack of depth d is located at a distance x_c from the left end. As before, there may be excitation terms to account for rigid body motion and other external excitation. In addition, since a fatigue crack applies a point bending moment, the spatial component of the nonlinear self-excitation is represented by a spatial doublet pair, or the second spatial derivative of the Dirac delta function at x_c. The unit nonlinear forcing function is then given in Equation (18) where $F_c(t)$ is the half-wave rectified sinusoid.

$$f_c(x,t) = F_c(t)\frac{d^2}{dx^2}\left(\delta(x-x_c)\right) \tag{18}$$

After expanding the general governing equation, the modal equations become

$$\begin{aligned} \ddot{T}_n(t) + 2\zeta_n\omega_n\dot{T}_n(t) + \omega_n^2 T_n(t) &= \int_0^L \left(\rho A f_\ell + \epsilon f_c(x,t)\right) W_n(x)dx \qquad (19) \\ &= f_\ell \int_0^L \rho A W_n(x)dx + \epsilon F_c(t)\int_0^L W_n(x)\frac{d^2}{dx^2}(\delta(x-x_c))dx \end{aligned}$$

To be concise, let Γ_n represent the definite integral of $\rho AW_n(x)$ over the length of the beam and apply integration by parts to the other integral in Equation (19). The result is a single set of closed-form modal equations given by

$$\ddot{T}_n(t) + 2\zeta_n\omega_n\dot{T}_n(t) + \omega_n^2 T_n(t) = \Gamma_n f_\ell(t) + \epsilon W_n''(x_c)F_c(t) \tag{20}$$

In order to make use of the modal equations in Equation (20), a critical step remains. Throughout the derivation, amplitude ϵ of the nonlinear forcing function has been left undefined with the only constraint being that it is not a function of position x since it was factored out of the spatial integral in Equation (19). However, it is likely proportional to the crack location x_c and the relative crack depth d, so a function of this form is considered.

Recalling the torsional spring approach, a crack behaves like a torsional spring that applies a point bending moment at the crack tip when the crack is open. The shape factor from Equation (??) is a polynomial in relative crack depth that can be used in the nonlinear amplitude. In addition, the curvature sign and magnitude at the location of the crack will affect the nonlinear forcing function and will be included in ϵ. Therefore, the following function will be considered for the amplitude of the nonlinear forcing function.

$$\epsilon = \alpha(x_c)F_1(d) \tag{21}$$

where $\alpha(x_c)$ is the amplitude of the ODS curvature at the crack location, and $F_1(d)$ is the depth-dependent shape factor.

For the spring hinge-free configuration, the shaker induces rigid body motion. Since there are no other external inputs, $F_{ext} = 0$. Restating the boundary conditions corresponding to the springhinge-free configuration:

$$\text{at} \quad x = 0: \qquad W_n(0) = 0, \qquad EIW_n''(0) = K_c W_n'(0) \tag{22}$$

$$\text{at} \quad x = L: \qquad W_n''(L) = 0, \qquad W_n'''(L) = 0 \tag{23}$$

Applying these boundary conditions to Equation (??) produces a system of equations that can be represented in matrix form as shown:

$$C\underline{a_n} = \begin{bmatrix} 1 & 0 & 1 & 1 \\ -\frac{\beta_n EI}{K_c} & -1 & \frac{\beta_n EI}{K_c}+1 & \frac{\beta_n EI}{K_c}-1 \\ -\cos\beta_n L & -\sin\beta_n L & e^{-\beta_n L} & e^{\beta_n L} \\ \sin\beta_n L & -\cos\beta_n L & -e^{-\beta_n L} & e^{\beta_n L} \end{bmatrix} \begin{bmatrix} a_{1n} \\ a_{2n} \\ a_{3n} \\ a_{4n} \end{bmatrix} = \begin{bmatrix} 0 \\ 0 \\ 0 \\ 0 \end{bmatrix} \tag{24}$$

For a nontrivial solution, the determinant of the coefficient matrix C must be zero. This yields the characteristic equation given in Equation (25) used to solve for β_n values that satisfy this constraint.

$$\frac{\beta_n EI}{K_c}(\cosh\beta_n L \sin\beta_n L - \cos\beta_n L \sinh\beta_n L) - \cos\beta_n L \cosh\beta_n L = 1 \tag{25}$$

Using the β_n values obtained from Equation (25) ensures $|C| = 0$, and also that the rows of C are not linearly independent. This in turn requires coefficients a_{1n}, a_{2n} and a_{3n} to be represented in terms of a_{4n}. To accomplish this, the fourth row of C is dropped as shown in Equation (26), and the remaining matrix is converted to reduced row echelon form as shown in Equation (27).

$$\begin{bmatrix} 1 & 0 & 1 & 1 \\ -\frac{\beta_n EI}{K_c} & -1 & \frac{\beta_n EI}{K_c}+1 & \frac{\beta_n EI}{K_c}-1 \\ -\cos\beta_n L & -\sin\beta_n L & e^{-\beta_n L} & e^{\beta_n L} \end{bmatrix} \begin{bmatrix} a_{1n} \\ a_{2n} \\ a_{3n} \\ a_{4n} \end{bmatrix} = \begin{bmatrix} 0 \\ 0 \\ 0 \end{bmatrix} \tag{26}$$

$$\text{reduces to:} \quad \begin{bmatrix} 1 & 0 & 0 & -c_{1n} \\ 0 & 1 & 0 & -c_{2n} \\ 0 & 0 & 1 & -c_{3n} \end{bmatrix} \begin{bmatrix} a_{1n} \\ a_{2n} \\ a_{3n} \\ a_{4n} \end{bmatrix} = \begin{bmatrix} 0 \\ 0 \\ 0 \end{bmatrix} \tag{27}$$

New coefficients c_{1n}, c_{2n} and c_{3n} in Equation (27) are functions of β_n, E, I, K_c and L and relate a_{1n}, a_{2n} and a_{3n} to a_{4n}. Using these relationships with Equation (??) yields

$$W_n(x) = a_{4n}\left(c_{1n}\cos\beta_n x + c_{2n}\sin\beta_n x + c_{3n}e^{-\beta_n x} + e^{\beta_n x}\right) \tag{28}$$

where a_{4n}, c_{1n}, c_{2n} and c_{3n} can have different values for each mode.

Fatigue Crack Identification

To make the response data useful for identifying fatigue cracks with either of the methods summarized, the ratio of fundamental and second harmonic magnitudes in the ASD of the response is used as the spectral signature for a given crack location and depth with excitation at a specified driving frequency. The spectral signature profile for a given mode is a collection of all crack signatures for a range of crack depths and locations relative to the healthy or zero signature. The zero signature is the spectral signature when crack depth goes to zero. Subtracting this value from the damaged spectral signatures yields the spectral signature profile.

The driving frequency for all spectral signatures within a profile will be half the natural frequency for the corresponding mode to ensure the largest harmonic resulting from the presence of a crack is amplified. Crack signature profiles must be generated for at leasts two modes before identification is possible for cases with an arbitrary crack of unknown depth and location as the two profiles must be cross-referenced. In cases where multiple predictions occur, additional modes are required.

When a crack is identified, an arbitrary crack is placed in the closed-form model, and the system is excited at driving frequencies corresponding to the crack signature profiles that will be used for identification. The spectral signature is determined for each case and plotted as a line on each of the corresponding crack signature profiles. The intersections between the spectral signature line from the test case and the crack signature profile are possible combinations of depth and location that could produce the observed spectral signature. The intersection points from each profile can then be used to plot prediction curves for each mode on a separate figure. The point at which these prediction curves intersect represents the common prediction between profiles and identifies the crack location and depth. Optimization can be used to ensure the fewest number of modes are used to identify the crack location and depth.

Figure (4) presents a comparison of the signature profiles for both methods for a 10% relative crack depth for modes 2-5 [14].

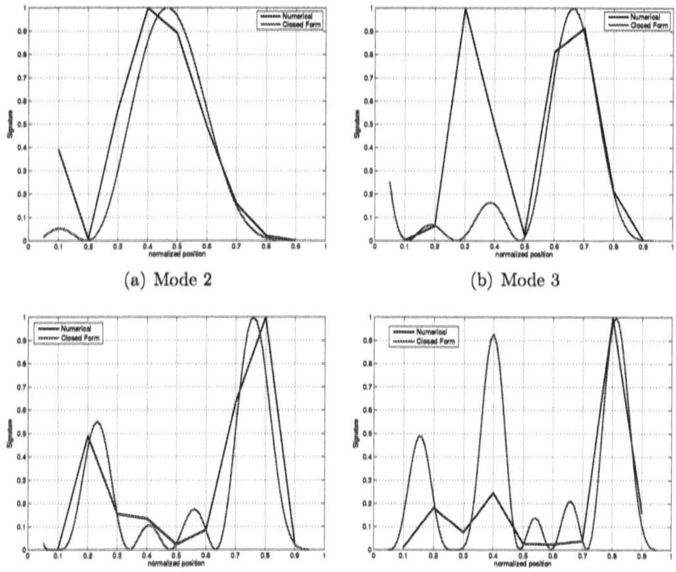

(a) Mode 2

(b) Mode 3

(c) Mode 4

(d) Mode 5

Figure 4: Signature profile comparison from previous work for each mode

Further Investigation of the Proposed Method

The closed-form approach served to reduce computation time by four orders of magnitude, but was only effective over a limited range of the structure. In addition, the tools used for crack identification were calibrated in an inconvenient manner as it required multiple cases to be run with the torsional spring model. Pinned-pinned boundary conditions are also considered to provide insight regarding the causes of the limited effective range observed in previous work[14]. Efforts to address these concerns are discussed in the results.

Pinned-Pinned Configuration

The pinned-pinned case was found in literature and served as the inspiration for much of this work. Since the closed-form method had a limited effective range when applied to a beam with a spring hinge-free configuration, examining the pinned-pinned configuration may provide insight. All data used for comparison to the proposed method outlined here is obtained directly from the work performed by Sundermeyer et al [12]. In the pinned-pinned configuration, each end of the beam is pinned as indicated. External excitation is provided at two locations x_A and x_B as shown in Equation (29), and $f_\ell = 0$.

$$\begin{aligned} F_{ext}(x,t) &= f_A(x,t) + f_B(x,t) \\ &= F_A \sin(\Omega_A t)\delta(x - x_A) + F_B \sin(\Omega_B t)\delta(x - x_B) \end{aligned} \tag{29}$$

For the closed-crack case, the translation and moment must be zero at each boundary ($x = 0$ and $x = L$) as shown in Equations (30) and (31).

$$\text{at} \quad x = 0: \qquad Y_n(0) = 0,\ Y_n''(0) = 0 \tag{30}$$

$$\text{at} \quad x = L: \qquad Y_n(L) = 0,\ Y_n''(L) = 0 \tag{31}$$

When the crack is open, the coordinates x_1 and x_2 are defined as before and used in place of x. The boundary conditions for the open case are given in Equations (32) and (33), and the continuity equations are the same as for the spring hinge-free configuration.

$$\text{at} \quad x_1 = 0: \qquad Y_1(0) = 0,\ \frac{d^2}{dx_1^2}[Y_1(0)] = 0 \tag{32}$$

$$\text{at} \quad x_2 = 0: \qquad Y_2(0) = 0,\ \frac{d^2}{dx_2^2}[Y_2(0)] = 0 \tag{33}$$

For the proposed closed-form method, a single model is used. The mass normalized mode shapes and natural frequencies are well known and are given in Equation (34) and (35).

$$W_n(x) = \sqrt{\frac{2}{\rho A L}} \sin\left(\frac{n\pi x}{L}\right) \tag{34}$$

$$\omega_n = \left(\frac{n\pi}{L}\right)^2 \sqrt{\frac{EI}{\rho A}} \tag{35}$$

As before, $f_\ell = 0$ and external excitation is provided at two locations x_A and x_B and is described by Equation (29). However, with this method the frequency must be determined for the Fourier series in $F_c(t)$. Since the nonlinear peak in the response ASD is at the difference of input frequencies, the driving frequency for the self-excitation is assumed to have the same form as shown in Equation (36).

$$\Omega_{dr} = \Omega_B - \Omega_A \tag{36}$$

Results

In order to better compare the spectral signature profiles between the two methods, a higher resolution data set is obtained for the torsional spring model at ten percent crack depth for each of the modes of interest 2-5. Whereas previously a crack was evaluated at every $0.1L$, the profiles in Figure (5) consider a crack at every $0.01L$. Upon examining the higher resolution signature profiles for the torsion spring method, many features are observed that were not previously present. In several instances, the curves are irregular when compared to the general shape of the closed-form profiles. Most interesting is the second mode where there are significant discontinuities.

The next consideration is an improved method for calibrating the closed-form signature profiles. In order for the signature profiles to be useful in crack identification, both the magnitude of the profiles and any test data obtained for identification must be scaled to a common range. As mentioned earlier, initial calibration was achieved with a ratio of maximum values from the profile for each method, which clearly defeats the benefits of the closed-form method. Various alternatives were tried, but the simplest and most effective solution found was to compare the same point in both profiles. Since only a single signature is compared, only one signature needs to be generated for each mode with the slower torsional spring method. While testing this method, it was found that the point chosen for comparison changed the relative magnitudes of the signature profiles, with the best results occuring where the closed from signature profiles were maximum. In consideration of this, the scale factor is now the ratio of the signatures in both profiles that correspond to the maximum value in the closed-form profile. First, the signature profiles for the closed-form method are produced and the maximum signature value is located. The crack location and depth corresponding to the maximum are identified, and a test is run with the torsional spring simulation at these settings to obtain the signature. With each of these values, the new scale factor is obtained but without the need to run long and costly simulations with the slower method. The results are provided in Table (1) and Figure (5).

Table 1: Scale factors for the spring hinge-free beam

Mode	Natural Frequency (Hz)	Driving Frequency (Hz)	Zero Signature (10^{-6})	Scale Factor
2	142	71	5.309	0.2364
3	400	200	0.0667	0.3532
4	786	393	0.0173	573.5
5	1306	653	0.0063	790.7

Although discrepancies are still observed, it is clear that the scale factor is just as effective for comparing the signature profiles. In addition, using the higher resolution torsion spring profiles makes it easier to determine which regions actually have poor coherence.

Finally, the pinned-pinned case is considered to evaluate what role the boundary conditions may have in the disparities observed. All settings are chosen to allow for direct comparison with Sundermeyer et al, and Figure (6) shows the ASD of a pinned-pinned beam with a crack at the midpoint. The second input frequency, Ω_B, is selected to ensure the difference frequency coincides with the first resonance. As expected, the nonlinear peak is amplified.

The primary difference between the results presented by Sundermeyer et al and the spectral signature profile in Figure 6(b) is symmetry. Although the profile produced with the torsion spring method was skewed and had a minor discontinuity, the profile produced with the closed-form method is continuous and relatively symmetric.

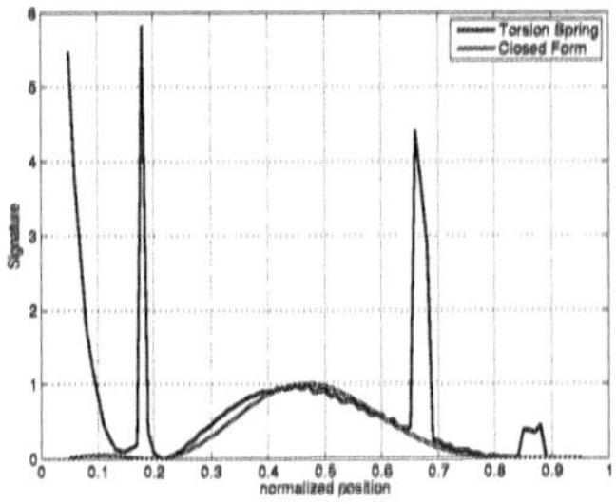

(a) Mode 2

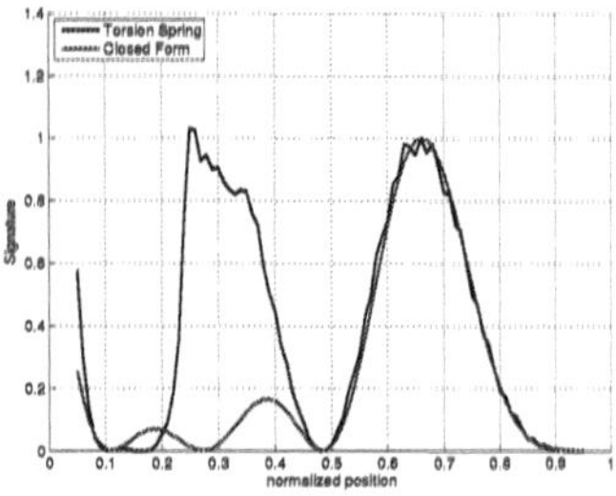

(b) Mode 3

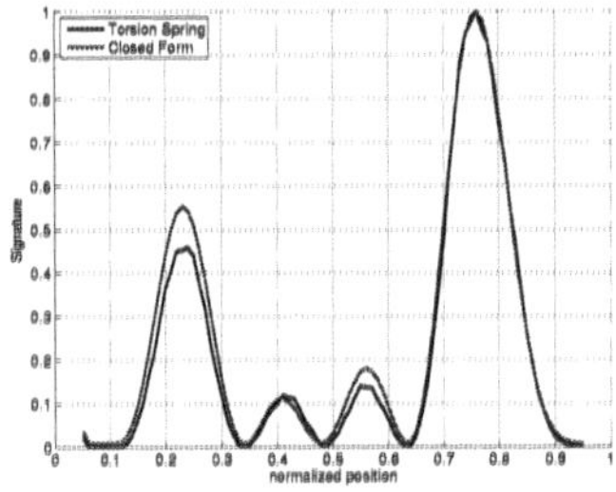

(c) Mode 4

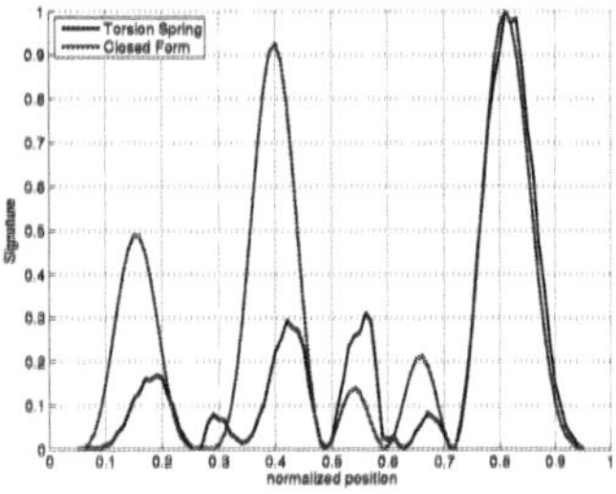

(d) Mode 5

Figure 5: Signature profile comparison for each mode

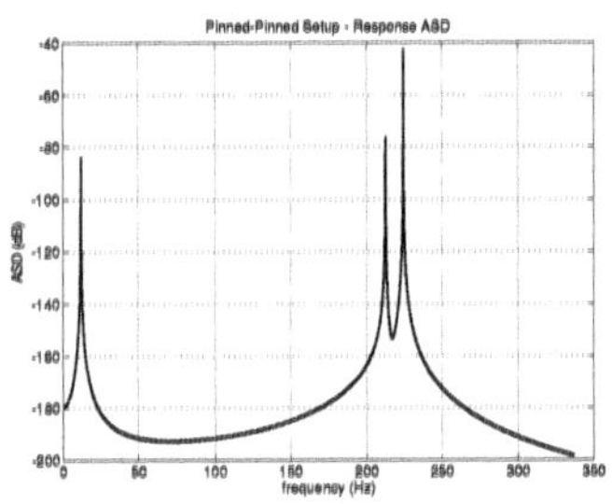

(a) ASD

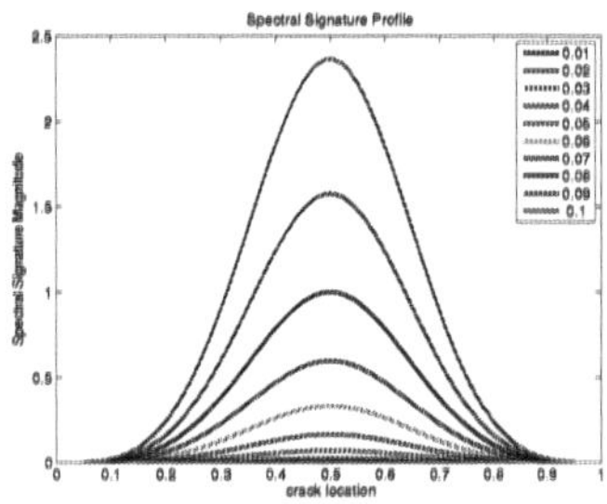

(b) Spectral Signature Profile

Figure 6: Results for the pinned-pinned closed-form solution.

Conclusion

In this research, additional investigation was performed to improve upon a closed-form method for obtaining the nonlinear response of a fatigue cracked beam. The method used for scaling and comparing the spectral signature profiles was examined and successfully altered to ensure time savings gained by the closed-form method are not forfeited. The number of crack locations considered with the torsion spring method was increased by a factor of ten, which provided considerably more detail for comparing the spectral profiles.

New features were observed that were previously unknown as they fell between samples. It is now easier to identify regions of disparity that must be rectified.

The closed-form method was applied to a pinned-pinned beam and compared to previously published work by Sundermeyer et al. The response ASD proved similar with a nonlinear peak appearing as expected, though the spectral signatures did not match. The discrepancies with this model will serve as an additional resource to help determine what potential changes can be made to the closed-form method to better approximate the nonlinear response of a cracked beam. The issues observed with the pinned-pinned configuration also imply that the problems observed with the spring hinge-free configuration are not solely related to the boundary conditions.

The next phase of investigation is to setup an experiment for obtaining the actual response of a cracked beam. Work has already begun on this effort, and the data obtained will be used for validating and modifying the closed-form method as necessary.

Appendix

Coefficients $\mathcal{A}_n$ and $\mathcal{B}_n$ in Eq. (14, 15, 16are expressed as

$$\mathcal{A}_n = \frac{d_{34}d_{12}d_{23} - d_{34}d_{13}d_{22} - d_{32}d_{23}d_{14} + d_{32}d_{24}d_{13} + d_{33}d_{22}d_{14} - d_{33}d_{24}d_{12}}{d_{33}d_{21}d_{14} - d_{33}d_{24}d_{11} - d_{31}d_{23}d_{14} + d_{31}d_{24}d_{13} + d_{34}d_{11}d_{23} - d_{34}d_{13}d_{21}} \tag{37}$$

$$\mathcal{B}_n = \frac{d_{31}d_{24}d_{12} - d_{31}d_{22}d_{14} + d_{34}d_{22}d_{11} - d_{34}d_{21}d_{12} - d_{32}d_{11}d_{24} + d_{32}d_{14}d_{21}}{-d_{32}d_{11}d_{23} + d_{32}d_{13}d_{21} + d_{33}d_{22}d_{11} - d_{33}d_{21}d_{12} - d_{31}d_{22}d_{13} + d_{31}d_{23}d_{12}} \tag{38}$$

where coefficients d_{mn} for each mode i are given as:

$$d_{11} = \left(1 - \frac{2\beta_n EI}{K_c}\right)\sin(\beta_n X_c) + \cos(\beta_n X_c) - e^{\beta_n X_c} \tag{39}$$

$$d_{12} = 2\sin(\beta_n X_c) + e^{-\beta_n X_c} - e^{\beta_n X_c} \tag{40}$$

$$d_{13} = -\cos(\beta_n(L - X_c)) - \sin(\beta_n(L - X_c)) - e^{\beta_n(L-X_c)} \tag{41}$$

$$d_{14} = 2\sin(\beta_n(L - X_c)) + e^{-\beta_n(L-X_c)} - e^{\beta_n(L-X_c)} \tag{42}$$

$$d_{21} = \sin(\beta_n X_c) - \left(1 - \frac{2\beta_n EI}{K_c}\right)\cos(\beta_n X_c) - e^{\beta_n X_c} \tag{43}$$

$$d_{22} = -2\cos(\beta_n X_c) - e^{-\beta_n X_c} - e^{\beta_n X_c} \tag{44}$$

$$d_{23} = \sin(\beta_n(L - X_c)) - \cos(\beta_n(L - X_c)) + e^{\beta_n(L-X_c)} \tag{45}$$

$$d_{24} = 2\cos(\beta_n(L - X_c)) - e^{-\beta_n(L-X_c)} - e^{\beta_n(L-X_c)} \tag{46}$$

$$\begin{aligned} d_{31} = -\cos(\beta_n X_e) \quad &- \quad \frac{K_t}{\beta_n EI}\sin(\beta_n X_c) - \left(1 + \frac{K_t}{\beta_n EI}\right)e^{\beta_n X_c} \\ &+ \quad \left(1 - \frac{2\beta_n EI}{K_c}\right)\left(\frac{K_t}{\beta_n EI}\cos(\beta_n X_c) - \sin(\beta_n X_c)\right) \end{aligned} \tag{47}$$

$$d_{32} = \left(1 - \frac{K_t}{\beta_n EI}\right)e^{-\beta_n X_e} - \left(1 + \frac{K_t}{\beta_n EI}\right)e^{\beta_n X_c} + 2\left(\frac{K_t}{\beta_n EI}\cos(\beta_n X_c) - \sin(\beta_n X_c)\right) \tag{48}$$

$$d_{33} = \frac{K_t}{\beta_n EI}\left(e^{\beta_n(L-X_c)} - \sin(\beta_n(L-X_c)) + \cos(\beta_n(L-X_c))\right) \quad (49)$$

$$d_{34} = \frac{K_t}{\beta_n EI}\left(-e^{-\beta_n(L-X_c)} - e^{\beta_n(L-X_c)} - 2\cos(\beta_n(L-X_c))\right) \quad (50)$$

Acknowledgments

The authors would like to thank Jeff Brown at Wright Patterson AFRL for sponsoring this topic, and the Dayton Area Graduate Studies Institute for continued funding with the award of a Ph.D. fellowship.

References

[1] "Safety Recommendation," Tech. Rep. A-10-98 to A-10-101, NTSB, May 2010.

[2] "Aircraft Accident Report - United Airlines Flight 232," Tech. Rep. AAR-90-06, NTSB, July 1989.

[3] Cowles, B., "High Cycle Fatigue in Aircraft Gas Turbines - An Industry Perspective," *International Journal of Fracture*, Vol. 80, July 1996, pp. 147–163.

[4] Gounaris, G. and Dimarogonas, A., "A Finite Element of a Cracked Prismatic Beam for Structural Analysis," *Computers and Structures*, Vol. 28, No. 3, 1988, pp. 309–313.

[5] Shiryayev, O. V. and Slater, J. C., "Improved Structural Health Monitoring Using Random Decrement Signatures: Application to FEM Data," *Structural Control and Health Monitoring*, Vol. 15, No. 7, November 2008, pp. 1006–1020.

[6] Shiryayev, O. V. and Slater, J. C., "Structural Damage Identification Using Random Decrement Signatures From Experimental Data," *Proceedings of the* 49th *AIAA/ASME/ASCE/AHS/ASC Structures, Structural Dynamics, and Materials Conference*, No. AIAA-2008-2163, April 2008.

[7] Shiryayev, O. V. and Slater, J. C., "Detection of Fatigue Cracks Using Random Decrement Signatures," *Structural Health Monitoring-An International Journal*, Vol. 9, No. 4, July 2010, pp. 347–360.

[8] Chati, M., Rand, R., and Mukherjee, S., "Modal Analysis of a Cracked Beam," *Journal of Sound and Vibration*, Vol. 207, No. 2, 1997, pp. 249–270.

[9] Andreaus, U., Casini, P., and Vestroni, F., "Non-linear Dynamics of a Cracked Cantilever Beam Under Harmonic Excitation," *International Journal of Non-Linear Mechanics*, Vol. 42, No. 3, 2007, pp. 566–575.

[10] Chatterjee, A., "Structural Damage Assessment in a Cantilever Beam with a Breathing Crack Using Higher Order Frequency Response Functions," *Journal of Sound and Vibration*, Vol. 329, No. 16, 2010, pp. 3325–3334.

[11] Andreaus, U. and Baragatti, P., "Cracked Beam Identification by Numerically Analysing the Nonlinear Behavior of the Harmonically Forced Response," *Journal of Sound and Vibration*, Vol. 330, No. 4, 2011, pp. 721–742.

[12] Sundermeyer, J. and Weaver, R., "On Crack Identification and Characterization in a Beam by Non-linear Vibration Analysis," *Journal of Sound and Vibration*, Vol. 183, No. 5, 1995, pp. 857–871.

[13] Shiryayev, O. V. and Slater, J. C., "Sensitivity Studies of Nonlinear Vibration Features For Detection of Cracks in Turbomachinery Components," *Proceedings of the* 51st *AIAA/ASME/ASCE/AHS/ASC Structures, Structural Dynamics, and Materials Conference*, No. AIAA-2010-3030, April 2010.

[14] Cooley, P. E., Slater, J. C., and Shiryayev, O. V., "Fatigue Crack Modeling and Analysis in Beams," *Proceedings of the* 53rd *AIAA/ASME/ASC/AHS/ASC Structures, Structural Dynamics, and Materials Conference*, No. AIAA-2012-1874, April 2012.

[15] Nayfeh, A. and Mook, D., *Nonlinear Oscillations*, Wiley Interscience, 1995.

ADVANCED MATERIALS I

Fatigue of Materials II Advances and Emergences in Understanding
Edited by: T.S. Srivatsan, M. Ashraf Imam, and R. Srinivasan
TMS (The Minerals, Metals & Materials Society), 2013

STRESS-CORROSION CRACKING AND FATIGUE CRACK GROWTH BEHAVIOR OF TI-6AL-4V PLATES CONSOLIDATED FROM LOW COST POWDERS

M. Ashraf Imam, Peter S. Pao and Robert A. Bayles

Materials Science and Component Technology Directorate, Naval Research Laboratory
Washington DC 20375, USA

Abstract:
Titanium is highly desirable for a wide range of applications because of its combination of high strength, low density and outstanding corrosion characteristics. However, the cost of titanium, produced by conventional technology, is high compared to steel and aluminum, which is a result of high extraction and processing costs. New approaches are being investigated maintaining required quality while lowering the cost of finished products. Ti alloy powder, Ti-6Al-4V, manufactured by a low cost hydride-process and consolidated into flat products (sheet, plate), were studied. The results of the study were compared with the properties obtained from plates of Armstrong Titanium consolidated powder. To remove the prior history of consolidation, the plates are beta annealed and the test results are compared with "as received" condition. The mechanism of the fatigue crack growth rate difference, fracture toughness, and stress-corrosion cracking resistance in terms of the respective Ti-6Al-4V microstructure differences will be discussed.

Key words: Titanium alloy, Armstrong-process, hydride-process, stress-corrosion cracking, fatigue crack growth.

1. Introduction

Titanium and its alloys are important for wide range of applications because of their lightweight, high specific strength, shock resistance, non-magnetic signature, and corrosion resistance characteristics. The high specific strength means that much lighter weight equivalent structures can be designed on ships enabling increased payload, range and speed. The corrosion resistance of titanium provides reduced maintenance and life cycle costs, along with reduced signatures through elimination of corrosion currents. The high melting temperature of titanium provides high tolerance to fire damage. The nonmagnetic properties of titanium provide increased survivability through reduced magnetic signatures, while also making titanium ideal for electronic component or mine countermeasure applications. However, because of its relatively high price – a result of extraction and processing costs- it is used basically only when it is the only choice; with the caveat that titanium has a bright "image" which can lead to use even when the economics are unfavorable. Broadly speaking, cost reduction can come from either reduction in the cost of production of the metal itself or from creative techniques for the fabrication of reliable components. Over the past few years there has been a lot of activity in the area of reduced cost titanium extraction processes. This is particularly a result of new and developing applications such as armor and auto use where a cost reduction could significantly increase use. However, it must be kept in mind that in the big picture the cost of extraction is a small fraction of the total cost needed to develop high-performance fabricated component.

Many new processes for reducing the cost of titanium have been discovered and researched over the last ten years. Several new companies have been founded to investigate and scale up these processes with the goal of reducing the cost of parts made from titanium metal and alloys. Two examples are ADMA Products Inc. (hydride process) and International Titanium Powders (ITP-Armstrong process). Many of the products of these new companies are in powder or pellet forms. Two processing streams for these low-cost materials can be envisaged: sponge substitute and direct consolidation. Simply using these new materials as feedstock for the old processing methods may produce significant savings. It is probable that greater savings, perhaps even revolutionary savings, may result from the combination of low-cost titanium feedstock and efficient processing to produce directly consolidated forms. The conventional method of consolidating the powder is to heat the compacted powder in a resistively heated or oil/gas-fired furnace. However, the time at temperature for consolidation is necessarily long because of the thermal inertia of the furnace leading to large grain size that in turn reduces the strength of the material. Evaluation of the mechanical properties of the consolidated product is important.

2. Experimental

Ti-6Al-4V Plates of approximately 13 mm thick with oxygen contents varying between 0.15 and 0.286 wt% (~ 0.44 to 0.85 at%) were used in this study. The plates were obtained from two different sources: one from ADMA Products Inc and the other from Concurrent Technology Corporation (Ti-6Al-4V powder processed at ITP). All the plates are consolidated at ADMA Products Inc. as shown in Figure 1. Table 1 shows range of chemical composition of the Ti-6Al-4V plates used. It is noted that the main difference in chemical composition between the plates are the oxygen content as noted in table 1. Hydrogen is removed during processing and the plates contain hydrogen less than 0.0054 wt% as noted in table 1.

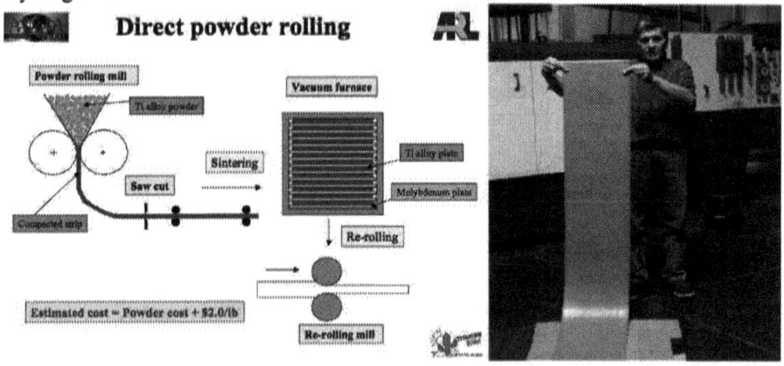

Figure 1. Direct Consolidation - a key to cost reduction (Courtesy of ADMA Products Inc.)

Table 1: Range of Chemical composition of the plates used.

Element, wt %	Ti-6Al-4V
Aluminum	5.88 - 6.16
Vanadium	4.21- 4.23
Iron	0.076 - 0.080
Oxygen	0.15- 0.286
Carbon	0.009 - 0.010
Nitrogen	0.008 - 0.021
Hydrogen	0.0018 - 0.0054
Yttrium	< 0.0005
Silicon	0.015-0.020
Chromium	0.020 - 0.028
Total Others	< 0.40
Other elements each	< 0.10
Titanium	Remainder

Optical micrographs of the as-received plates are shown in Figure 2. It is noted that the plate from CTC with $O_2 = 0.208$ wt% shows microstructure (Figure 2a) similar to beta annealed. The plate, HD 64A-5, with $O_2 = 0.286$ wt% shows alpha-beta annealed microstructure (Figure 2b) whereas the plate, HD 64D, with $O_2 = 0.15$ wt% also shows alpha-beta annealed microstructure (Figure 2c) with texture compared to HD 64A-5 (Figure 2b) plate.

The samples were tested in two different conditions: one set of samples in as-received and the other set as beta annealed. Beta annealing was done by heating the samples in vacuum to 1038 °C and holding at the temperatures for 0.5 hr and then cooled to room temperature in argon. This is followed by annealing at 732 °C for 2 h in vacuum, then cooled to room temperature in argon. The microstructures of these samples are shown in Figure 3. The goal of the beta annealing was to remove prior history of the plates of Ti-6Al-4V and produce the same prior beta grain size and Widmanstatten packet size. This microstructure was not achieved for sample (HD 64A-5) with $O_2 = 0.286$ wt% as shown in Figure 3b where the remnants of alpha phase are evident. The average prior beta grain sizes of these samples after annealing are also different and range from 0.15 to 0.5 mm. Since oxygen contents of these plates are different, the beta transus temperatures of samples made from these plates are different [1]. The beta annealing of these samples was done under the same condition with the understanding that all samples have the beta transus temperature less than 1038°C. This was not the case for the sample (HD 64A-5) with $O_2 = 0.286$ wt% as evidenced by microstructure in Figure 3b.

Compact tension specimens, having 25.4-mm and 10.2-mm thickness, were used, respectively, for fracture toughness determination, for Stress-Corrosion Cracking (SCC), and fatigue threshold studies. To enhance constraint so as to more closely simulate the more conservative plane strain condition and to increase the chances of cracking in the desired plane, a 5% side groove (90° root angle and 0.254 mm maximum root radius) was introduced on both sides of each compact tension specimen. For SCC studies, rising-step-load tests were used to determine the sub-critical cracking thresholds under SCC conditions in 3.5% NaCl solution [2]. Before rising-step-load testing, the specimens were fatigue pre-cracked using load shedding techniques to achieve final pre-cracking stress intensity less than 60% of the expected sub-critical cracking threshold. The load was applied by a screw-driven test machine with a horizontal load axis. This allows the CT specimen to be suspended in a dish of salt solution. A hold time of 15 minutes was used for each step load. The stress intensity loading rate during the rising load portion of each step was about 0.0167 $MPa\sqrt{m}$/s. Each load step was approximately 1.8 $MPa\sqrt{m}$ as shown in Figure 4. Crack growth is indicated by a downward slope in the load vs. time curve, so the threshold K value reported is computed from the load of the next lower step. Following the rising-step-load subcritical cracking tests, the fracture surfaces were examined by scanning electron microscopy (SEM) to identify the fracture mode and cracking mechanisms.

For the fatigue tests, the load was applied by a servo-hydraulic actuator with a horizontal load axis. As with the SCC test, this allows the CT specimen to be suspended in a dish of salt solution. The fatigue loads were applied at 10 Hz with R = 0.10. The loads were varied to generate the complete delta K range on a single specimen.

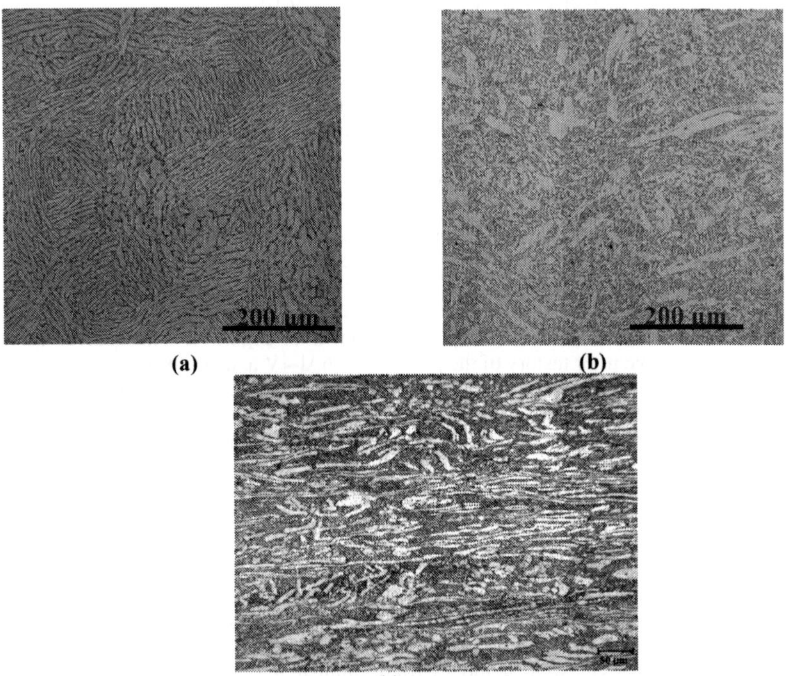

(c)

Figure 2. Optical micrographs of as received (a) CTC 64-5 with O_2 = 0.208 wt% (b) HD 64A-5 with O_2 = 0.286 wt% (c) HD 64D with O_2 = 0.15 wt%.

3. Results and Discussion

3.1. Fracture Toughness

The fracture toughness data presented in Table 2 for all the samples and plotted against oxygen content in Figure 5 for beta annealed specimens only do not satisfy the strict plane strain constraint requirements. Nevertheless, since all specimens are of the same thickness, we can still compare their fracture toughness. As shown in Figure 5 for beta annealed specimens, fracture toughness increases linearly with decreasing oxygen concentration. However, the as-received hydrided samples have low fracture toughness compared to beta annealed except the CTC samples which might have been beta annealed when we received the samples. It is interesting to note the effect of oxygen content on fracture toughness of beta annealed samples. There are other variables that could affect the fracture toughness such as grain size, chemical composition and texture. Even though there are variations in grain size and chemical composition of the beta annealed samples, the effect of oxygen seems more prevalent.

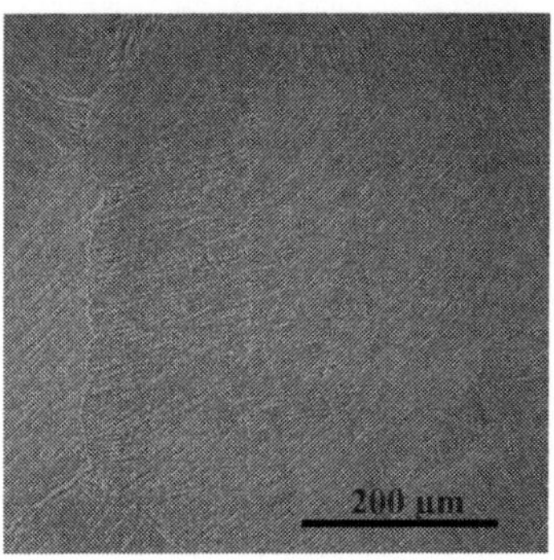

(a)

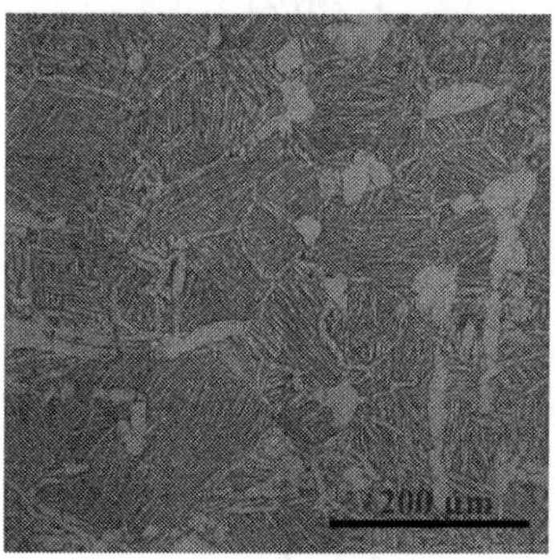

(b)

(c)

Figure 3. Optical micrographs of beta annealed (a) (a) CTC 64-5 with O_2 = 0.208 wt% (b) HD 64A-5 with O_2 = 0.286 wt% (c) HD 64D with O_2 = 0.15 wt%.

Table 2: Fracture toughness of Ti-6Al-4V as function of oxygen content and heat treatment.

Material Ti-6Al-4V; ~0.5" thick	**Specimen** (ID #)	**Oxygen** (Wt%)	**Heat Treatment**	K_{scc} **MPa√m** **(ksi√in)**, Average
ADMA	HD64D	0.15	As-Received*	22.0 (20.0)
ADMA	HD64D	0.15	Beta Annealed	66.4 (60.4)
ADMA	HD64A-5	0.286	As-Received*	23.7 (21.6)
ADMA	HD64A-5	0.286	Beta Annealed	34.6 (31.5)
CTC (ITP)	CTC64-5	0.208	As-Received**	66.0 (60.1)
CTC (ITP)	CTC64-5	0.208	Beta Annealed	56.8 (51.7)

* alpha-beta-anneal.

** microstructure shows beta-annealed in as-received condition.

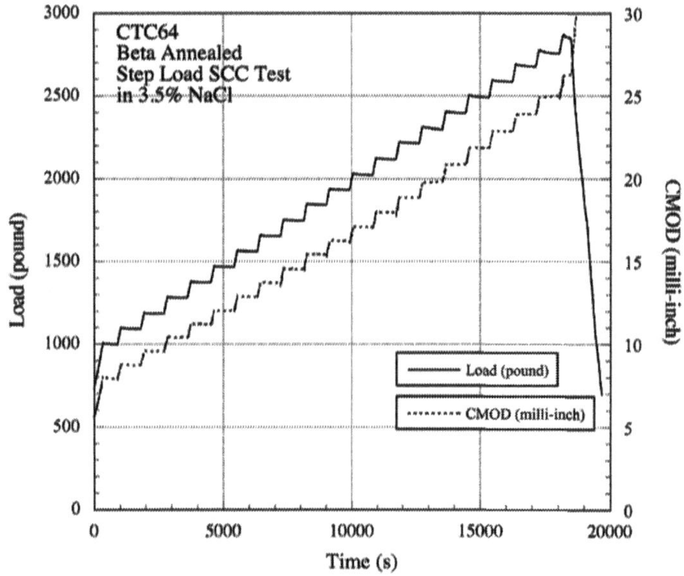

No apparent subcritical crack growth (Kscc = 56.9 MPa √m)

(For Typical beta annealed-- Kscc =60 MPa √m)

Figure 4. Rising Step Load SCC Test in 3.5% NaCl (Beta Annealed CTC64)

3.2. Stress-Corrosion Cracking

The SCC threshold stress intensities obtained in 3.5 wt% NaCl solution for all the plates of Ti-6Al-4V are shown in Figure 5. The SCC threshold data for a typical sample of Ti-6Al-4V (O_2= ~0.10) obtained from earlier investigations are also included in Figure 6 for comparison [3]. It is interesting to note that the SCC threshold of Ti-6Al-4V begins to decrease immediately as oxygen content starts to increase. However, as the oxygen concentration increases there is only a modest decrease in SCC threshold. The low oxygen-containing Ti-6Al-4V alloys, which exhibit high SCC thresholds, failed under sustained-load cracking (SLC). It is clear from Figure 5 that for good SCC resistance, keeping oxygen concentration low is important.

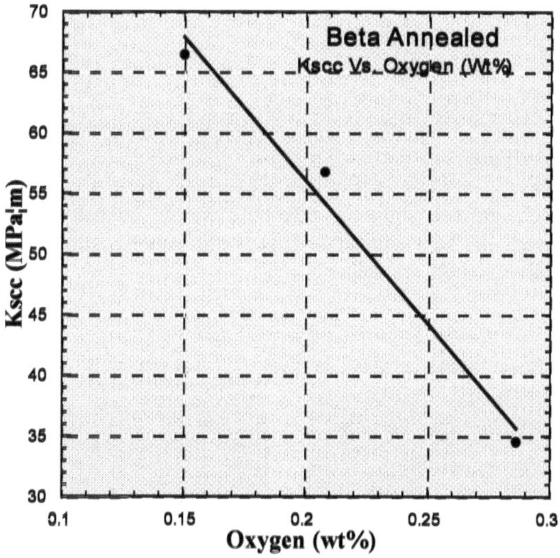

Figure 5. Stress intensity as a function of oxygen content for beta annealed Ti-6Al-4V.

3.3. Fatigue Crack Growth

The fatigue crack growth kinetics are obtained at R = 0.10 for Ti-6Al-4V with various oxygen concentration in an ambient air. The fatigue crack growth rate data for Ti-6Al-4V with oxygen contents ranging from 0.15 to 0.286 wt% all falls into a very narrow band. The fatigue crack growth rate at certain stress intensity factor range and the fatigue crack growth threshold stress intensity range are about the same for each Ti-6Al-4V alloy. Thus, oxygen levels in the range of 0.15 to 0.286 wt% have no effect on the fatigue crack growth kinetics of Ti-6Al-4V.

3.4. Fractographic Analyses of Fatigue Specimens

The representative fracture surface morphologies of as-received and beta annealed conditions of for two samples, HD64A and CTC64, are shown in Figures 7 to 10. Fracture surface morphologies of HD64D are not presented.

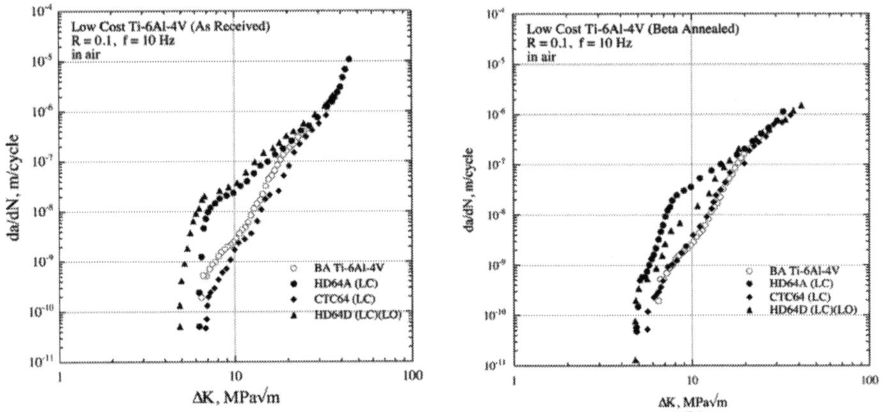

(a) (b)

Figure 6. Fatigue crack growth showing threshold stress intensity for different samples (a) as-received, (b) beta annealed. This shows more scatter in the threshold of as-received relative to beta annealed.

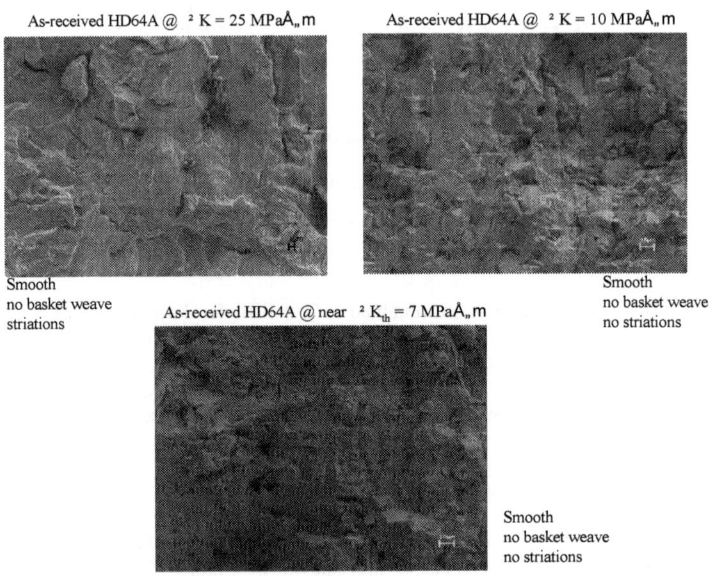

Figures 7. Fatigue fracture surface morphologies of as-received HDA64A

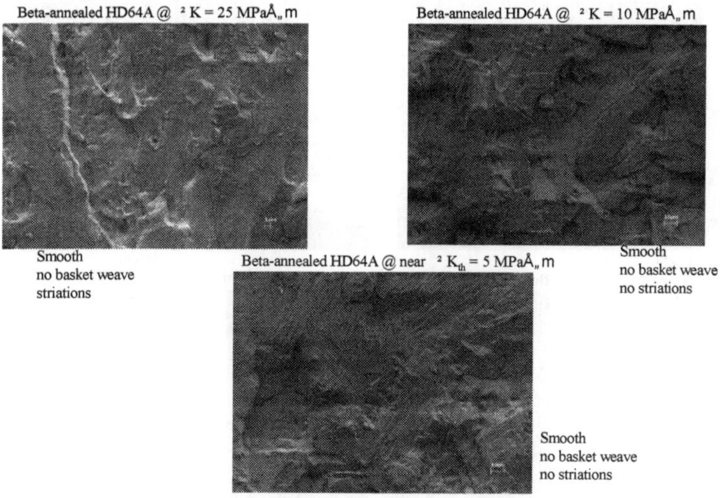

Figures 8. Fatigue fracture surface morphologies of beta annealed HDA64A

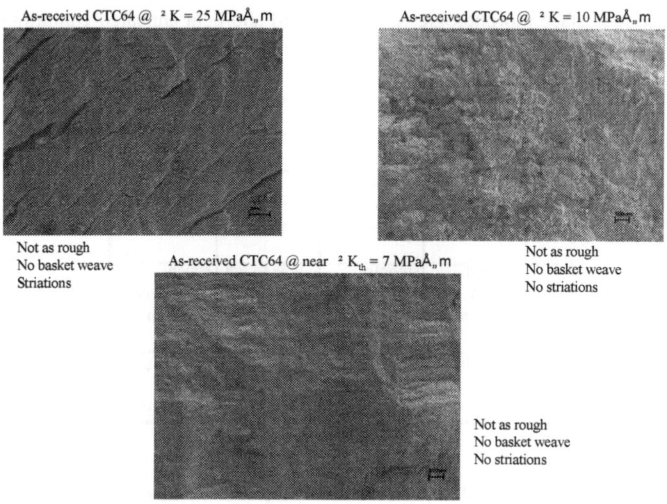

Figures 9. Fatigue fracture surface morphologies of as-received CTC64

Beta-annealed CTC64 @ ² K = 25 MPaÂ„m

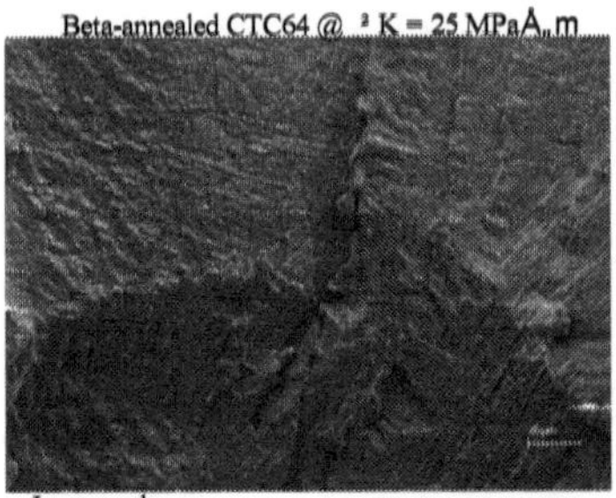

Less rough
Basket weave
Striations

Beta-annealed CTC64 @ ² K = 10 MPaÂ„m

Rough
Basket weave-not Clear
No striations

Beta-annealed CTC64 @ near ² K_{th} = 6 MPaÂ„m

Rough
Basket weave-not Clear
No striations

Figures 10. Fatigue fracture surface morphologies of beta annealed CTC64

Table 3. Fracture surface morphologies of as-received and beta annealed conditions of for two samples, HD64A and CTC64, as shown in Figures 7 to 10

² K MPaÂ„m	HD64A		CTC64	
	AR	BA	AR	BA
² K_{th}	smooth no BW no striations	smooth no BW no striations	not as rough BW not clear no striations	rough BW not clear no striations
10	smooth no BW no striations	smooth BW no striations	not as rough no BW no striations	rough BW not clear no striations
25	smooth no BW striations	smooth BW striations	not as rough no BW striations	less rough BW striations

As shown in Figure 7 the SCC fracture surface morphologies of the as received in Ti-6Al-4V with 0.286 wt% oxygen consists primarily of micro void coalescence and coarse irregular flutes.

The beta annealed Ti-6Al-4V with 0.286 wt% oxygen, Figure 8, has similar fracture surface morphology as that shown in Figure 7. This type of fracture surface morphology is also similar to the fast fracture region in air. Similarly, for Ti-6Al-4V containing 0.208 wt% oxygen, their fracture surface morphologies are about the same. Cleavage cracking, which is characteristic of many SCC fractures in titanium alloys, was not found in the SCC crack areas [4-7]. The fluting observed was not qualitatively similar to that seen in classic SCC fracture surfaces of susceptible higher-oxygen titanium alloys, being much less regular and straight-sided. Therefore, SCC was probably not the cause of the observed slow subcritical cracking in Ti-6-4. Rather, the subcritical cracking mechanism for these alloys in salt water environment is more likely to be similar to that responsible for inert-environment sustained-load cracking of titanium alloys [3-4].

As the oxygen concentration increases the fracture surface morphology is characterized by the presence of many regular flutes and river patterns. Flutes are ruptured halves of tubular voids that form on prismatic surfaces by planar, intersecting slip on {1010} planes in the alpha phase [4]. The tearing ridges that delineate the flutes are straight, merging into one another abruptly to form river patterns. Interstitial oxygen in titanium alloys promotes fluting by restricting cross slip and facilitating columnar slip [4]. SCC conditions can further promote flute formation in titanium alloys indirectly by causing multiple cleavages. The presence of characteristic regular flutes suggests SCC susceptibility in Ti-6Al-4V.

When Ti-6Al-4V contains high oxygen concentrations, the SCC fracture surface exhibits the flat and sharply defined appearance as shown in Figures 9 and 10. This type of cleavage fracture is often observed in salt water environments for which the SCC threshold is quite low. The fracture path is transgranular and trans-Widmanstatten packet. Because a Widmanstatten packet consists of similarly aligned α plates (separate by interplate β phase), SCC cracks can cleave on a common plane through all the α plates in a single packet, and hence, form large and flat cleavage facets. The cleavage occurs only in the α phase while the b phase does not cleave, but tears. The tearing ridges that separate the α plates can be seen. These fractographic observations are consistent with the low SCC threshold in other titanium alloys that contains high levels of internal oxygen and confirm their poor SCC resistance in salt water environments.

4. Summary

1. Oxygen plays an important role in mechanical properties but systematic studies of its role on Ti-6Al-4V show no significant effect on fatigue crack growth kinetics but have significant influence on SCC and K_{ssc}.
2. Fracture toughness is linearly related to oxygen content, higher oxygen content lowers the fracture toughness
3. CTC64 has significantly coarser grain size than that of HD64A.
4. In the low and intermediate ΔK regions, both as received and beta annealed CTC64 exhibit significantly lower fatigue crack growth rates when compared to HD64A.
5. The observed fatigue crack growth responses may be attributed to the significantly coarser grain size in CTC64, which produces crack deflection, closure and lower the crack driving force.
6. CTC64 has significantly higher apparent Kscc than HD64A in 3.5% NaCl. Rising step load SCC tests and post-fracture fractographic analyses do not suggest SCC sub-critical crack growth.

7. The minor difference in fracture toughness between two conditions of CTC64-5, as-received and beta annealed, may be related to different grain size difference caused by beta annealing since the microstructure seems very similar.
8. In ADMA plates, the as-received plates clearly show alpha–beta annealing condition with texture which may be the cause of lower fracture toughness.

Acknowledgement

This work was supported by the Office of Naval Research. The authors acknowledge the support of ADMA Products Inc ((V. S. Moxson and V.A. Duz ADMA) and Concurrent Technology Corporation (F. Robert Dax and Yuan Pang) for providing the plates of Ti- 6Al-4V.

REFERENCES

1) M.A. Imam, B.B. Rath, and D.J. Gillespie: *Proc. 5th Int. Conf. on Titanium*, ed. by G. Lutjering, U. Zwicker, and W. Bunk, (Deutche Gesellschaft fur Metallkunde e.V., Adenauerallee, 1985) pp. 1511-1518.

2) *G.R. Yoder, L.A. Cooley, and T.W. Crooker: Fatigue Crack Growth Measurement and Data Analysis, ASTM STP 738, ed. by S.J. Hudak, Jr. and R.J. Bucci, (ASTM, Philadelphia, 1981) pp. 85-102.*

3) *P.S. Pao, M.A. Imam, S.J. Gill, C.R. Feng, and R.A. Bayles: Proc. 10th Int. Conf. on Titanium, ed. by G. Lutjering and J. Albercht, (Wiley-VCH, Weinheim, 2004) pp. 2083-2090.*

4) D.A. Meyn and E.J. Brooks: *Fractography and Materials Science, ASTM STP 733*, ed. by L.N. Gilbertson and R.D. Zipp, (ASTM, Philadelphia, 1981) pp. 5-31.

5) D.A. Meyn: Met. Trans. 5, (1974) pp. 2405-2414.

6) D.A. Meyn and R.A. Bayles: *Fractography of Modern Engineering Materials: Composites and Metals, ASTM STP 948*, ed. by J.E. Masters and J.J. Au (ASTM, Philadelphia, 1987) pp. 400-423.

7) D.A. Meyn and P.S. Pao: *Slow Strain Rate Testing for the Evaluation of Environmentally Induced Cracking: Research and Engineering Applications, ASTM 1210*, ed. by R.D. Kane (ASTM, Philadelphia, 1993) pp. 158-169.

ADVANCED MATERIALS II

Fatigue of Materials II Advances and Emergences in Understanding
Edited by: T.S. Srivatsan, M. Ashraf Imam, and R. Srinivasan
TMS (The Minerals, Metals & Materials Society), 2013

THE HIGH CYCLE FATIGUE, DAMAGE INITIATION, DAMAGE PROPAGATION and FINAL FRACTURE BEHAVIOR of ALUMINUM ALLOY 2024

T.S. Srivatsan[1], Satish Vasudevan[2] and K. Manigandan[3]

1: ***Professor***
3: Graduate Student
Department of Mechanical Engineering
The University of Akron
Akron, Ohio 44325-3903, USA
E-Mail: tss1@uakron.edu

2: *Product Development Engineer*
[Custom Engineering]
Navistar, Inc.
2701 Navistar Drive, Lisle, IL 60532
E-Mail: Satish.Vasudevan@Navistar.com

ABSTRACT

In this technical paper the results of a study aimed at understanding the high cycle fatigue properties and fracture characteristics of aluminum alloy 2024 is presented and discussed. Specimens of the alloy in the T-8 temper were cyclically deformed over a range of stress amplitudes at ambient temperature and at a stress ratio of 0.1. Specimens of the alloy were taken from the longitudinal orientation of the as-provided plate and cyclically deformed. The influence of alloy temper (T8 versus T3) on cyclic fatigue life under stress amplitude control is briefly discussed. At the ambient test temperature, the macroscopic fracture mode was essentially identical with specific reference to the magnitude of cyclic stress amplitude. The microscopic mechanisms governing cyclic deformation, fatigue life and final fracture behavior are discussed in light of the mutually interactive influences of magnitude of applied stress, intrinsic microstructural effects, deformation characteristics of the alloy microstructure and macroscopic fracture mode.

Key Words: aluminum alloy, microstructure, cyclic fatigue, stress amplitude, fracture behavior

Introduction

Noticeable strides in both engineering and technology-related applications coupled with an increased emphasis and importance given in the aerospace industry, which successfully pushed technology to its limits, to one that is largely dictated by both commercial interests and concurrent need, did provide the much needed interest, inspiration, inspiration and impetus for developing new, improved and essentially low cost alloys as viable replacements to the material currently being used [1-3]. Both current and emerging technology requirements have placed an increased emphasis on issues like off-the shelf availability, lower weight, damage tolerance, durability, fabric ability, high life-cycle design and even cost effectiveness as primary consideration in governing not only the development of materials but also their selection and use [4-6]. During the past three decades more than one approach has been tried in developing alloys that can provide the desired properties at room temperature in addition to possessing strength at high temperatures and high temperature super-plastic formability. In particular, the alloys of aluminum having during the last four decades, since the early 1970's, engendered a combination of scientific and technological interest for use in a spectrum of high performance applications spanning the industries of aerospace, automotive and even commercial products. The growing need for aluminum alloys that offer a combination of good fracture toughness, high strength, adequate damage tolerance, good high cycle fatigue resistance and even acceptable corrosion resistance that will enable their use in airframe structures and army armaments led to sustained efforts aimed at the development of new and improved alloys, tempers and even their composite counterparts [3,6,7]. It has been safely predicted that most of the aluminum alloys and its advanced or emerging variations currently being produced will find use for those applications that did not exist three or four decades ago [6]. The present markets for aluminum and its alloys are greatly affected by shortages of the other metal counterparts coupled with a combination of economical, environmental and safety factors [7]. In more recent years, the technology related to both aluminum and alloys have grown stronger with competition, i.e., with an increased emphasis on both physical and mechanical properties of the alloys themselves and the dominant role they play in structural applications.

The domain of high cycle fatigue (HCF) corresponds to small applied stress that is essentially confined to the linear region of the stress versus strain curve obtained in a simple tension test. The resultant number of cycles-to-failure (N_f) is high and often greater than 10^4 cycles. It usually comprises of two distinct stages, which have been traditionally referred to as Stage I and Stage II [8]. During Stage I, for polycrystalline metallic materials without defects or inclusions, the cracks, both microscopic and macroscopic in size, appear along the slip bands in grains that are favorably oriented with respect the major stress axis and subsequently tend to follow the preferential slip systems in the adjacent or neighboring grains [8-10]. This stage is generally influenced by intrinsic microstructural features of the alloy. During Stage II the fine microscopic cracks tend to propagate easily along those planes having a maximum normal stress [8,9]. For a polycrystalline ductile solid that is subjected to high cycle fatigue, Stage I is expected to be much larger than Stage II and therefore the number of cycles to failure

(N_f) is strongly dependent on microstructure. The intrinsic microstructural features exert a profound influence on the following: (i) the kinetics of crack initiation, (ii) early microscopic and macroscopic crack growth through the alloy microstructure, (iii) ensuing stable crack growth through the microstructure, and (iv) crack arrest depending on orientation of the grain with respect to the far-field stress axis. This partially explains the scatter that is likely to occur at the fine microscopic level and is quantified by the number of cycles-to-failure [11].

The primary objective of this experimental study was to provide an insight into the high cycle fatigue and final fracture behavior of aluminum alloy 2024 in the T8 temper. The high cycle fatigue response and final fracture behavior are discussed in light of the mutually interactive influences of magnitude of maximum cyclic stress, intrinsic microstructural effects, matrix deformation characteristics and macroscopic aspects of fracture.

Material and Sample Preparation

The material chosen and used in this experimental study was rolled aluminum alloy 2024 plate provided by the Boeing Commercial Airplane Company. The as-provided plates had a thickness of 3-mm. The nominal chemical composition of the Al-Cu-Mg alloy is given in Table 1. Test specimens for the mechanical tests (tension and cyclic fatigue) were precision machined from the as-provided plate stock and conformed to specifications outlined in Standard ASM E-8-10 [Standard Method for Tensile Testing of Metallic Materials] [12] and ASTM E-466-10 [Standard test Method for Stress Amplitude Controlled High Cycle Fatigue Tests on Metallic Materials] [13]. All of the test specimens were precision machined from the as-provided plate stock such that the major stress axis was along the longitudinal direction of the plate. To minimize the extrinsic influence of surface irregularities and finish, and considering surface sensitiveness of the samples prepared from the as-provided material, abundance of care was taken during handling of the test specimens up until testing them on the servo-hydraulic mechanical test machine. To minimize the effects of surface irregularities and finish, final surface preparation of each machined test specimen was achieved by mechanical polishing the gage section using progressively finer grades of silicon carbide impregnated emery paper with the purpose of removing any and all circumferential scratches and surface machining marks.

Experimental Procedures

Initial Microstructure Characterization.

Samples for metallography observation were cut from the as-provided plate stock of alloy 2024-T8. The cut samples, measuring 25 mm square in cross-section, were mounted in bakelite and then wet ground on progressively finer grades of silicon carbide impregnated emery paper using copious amounts of water both as a lubricant and as a coolant. Subsequently, the ground samples were mechanically polished using five-micron and one-micron aluminum oxide (Al_2O_3) powder suspended in distilled water as the lubricant.

The polished aluminum alloy sample was etched using Keller's reagent (a solution mixture of hydrofluoric acid, concentrated nitric acid and distilled water). The etched surface of the sample was observed in an optical microscope and photographed using standard bright field illumination technique.

Mechanical Testing

All mechanical tests were performed up until failure on a fully-automated, closed-loop servo-hydraulic mechanical test machine (INSTRON) equipped with a 100 KN load cell. The specimens were deformed in uniaxial tension at a constant strain rate of 0.0001/sec. The tests were performed in ambient temperature laboratory air environment (Relative Humidity of 55 pct, Temperature = 25°C). Cyclic stress-amplitude controlled high cycle fatigue tests were conducted at ambient temperature (25°C) at a constant cyclic frequency of 5Hz and stress ratio [R = minimum stress/maximum stress = 0.1]. The stress and corresponding number of cycles-to-failure (N_f) were recorded on a PC-based data acquisition system.

The yield strength value obtained from the uniaxial tensile test was used to calculate the maximum stress for the cyclic fatigue tests. At the chosen stress ratio the fatigue tests were conducted over a range of stress amplitudes to establish the variation of maximum stress ($\sigma_{maximum}$) with fatigue life (N_f).

Failure-Damage Analysis

Fracture surfaces of the cyclically deformed and failed specimens were carefully examined in a scanning electron microscope (SEM) over a range of allowable magnifications to:

(a) Determine the macroscopic fracture mode.

(b) Characterize the fine scale topography and features on the fatigue fracture surface for the purpose of establishing (i) the nature of crack initiation, (ii) nature of crack propagation through the alloy microstructure, and (iii) other intrinsic mechanisms governing fracture at the fine microscopic level.

Samples for SEM observation were obtained from the failed fatigue specimens by sectioning parallel to the fatigue surface.

Results and Discussion

Initial Microstructure

The microstructure of the as-provided 2024-T8 alloy plate is shown in **Figure 1**. The microstructure was fully recrystallized with fairly large recrystallized grains that were flattened and elongated along the longitudinal direction as a direct consequence of the mechanical deformation induced by the rolling operation. All of the micrographs were taken immediately below the surface of the specimen chosen. The recrystallized grains were observed to be non-uniform in size and shape. The coarse second-phase particles were found to be randomly distributed through the alloy microstructure with clustering or agglomeration occurring at regular intervals. A careful examination of the polished sample surface revealed few of the coarse second-phase particles to be decorating the high angle grain boundaries. The smaller and finer second-phase particles, i.e., the Al_2Cu precipitates, could not be easily observed at the allowable magnifications of the optical microscope used in this study (< 1000X).

Tensile Properties

The ambient temperature tensile properties of the as-provided 2024-T8 plate are summarized in **Table 2**. The results reported are the mean value of duplicate tests. At the end of the test, the values of yield strength (σ_{YS}), ultimate tensile strength (σ_{UTS}), fracture strength ($\sigma_{fracture}$) and elongation-to-failure (ϵ_f) were provided as an output of the PC-based data acquisition system. The yield strength of the candidate alloy was determined by identifying the stress at the point on the engineering stress versus engineering strain curve where a straight line drawn parallel to the elastic portion of the curve, at 0.2 pct offset, intersects the curve. The ductility is reported as elongation-to-failure over a gage length of 12.7 mm (0.5 inch). The strain-to-failure was measured using a clip-on extensometer that was attached to the test specimen at the gage section. The test results reveal only marginal difference in both strength and ductility for the two most widely used tempers of this alloy, i.e., T3 and T8. The yield strength of this alloy is 445 MPa, while its ultimate tensile strength in the T8 temper is 476 MPa, suggesting the occurrence of low strain hardening beyond yield. This is clearly seen in the engineering stress versus engineering strain curve shown in **Figure 2**. The elongation measured over 12.5 mm gage length was 8.84 percent.

High Cycle Fatigue Response

The primary purpose of the stress amplitude-controlled fatigue tests is to establish the influence of maximum stress on fatigue life. At low values of the maximum stress the material can be cycled indefinitely, defined in this study as cycles-to-failure in excess of one-million (10^6) cycles. Many metals, to include the ferrous-based family, exhibit a well-defined endurance limit. The endurance fatigue limit is the stress below which the material does not fail. In this experimental study the tests were set to terminate at 10^6 cycles and if failure did not occur in this cycle period the chosen sample is taken to have an infinite life at this stress level. A difference in cyclic fatigue life between the test specimen deformed at a maximum stress near the endurance limit and a higher maximum stress is distinctly evident or noticeable. For most practical applications the material is assumed to have infinite life at and below the endurance limit.

Pure aluminum and its alloys are regarded as not having a plateau in the maximum stress versus fatigue life relationship, unlike the characteristics shown by other non-ferrous metal counterparts having BCC and HCP crystal structures and even the ferrous-based alloys. In this study, an arbitrary cut-off value of 10^6 cycles was used to establish the endurance limit. The results of the axial stress amplitude-controlled fatigue tests is shown in **Figure 3** in which the maximum stress ($\sigma_{maximum}$) is plotted as a function of cycles-to-failure (N_f). The maximum stress ($\sigma_{maximum}$) versus fatigue life (N_f) curve

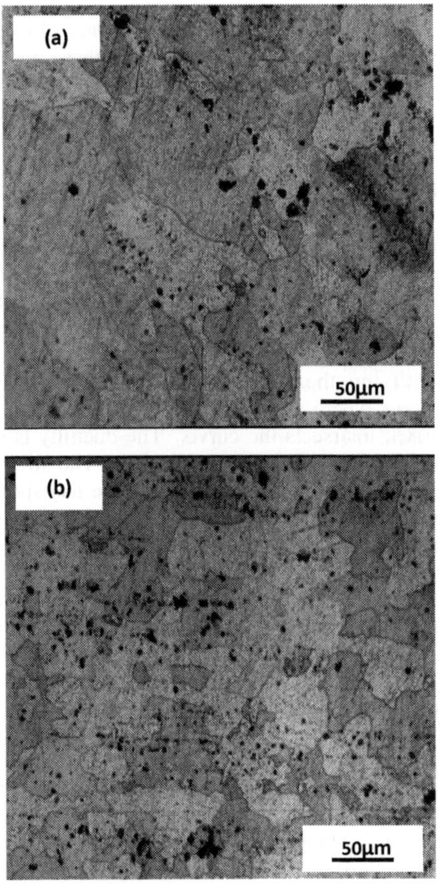

Figure 1. **Optical micrographs showing microstructure of aluminum alloy 2024-T8 (a) Longitudinal direction, and (b) Transverse direction**

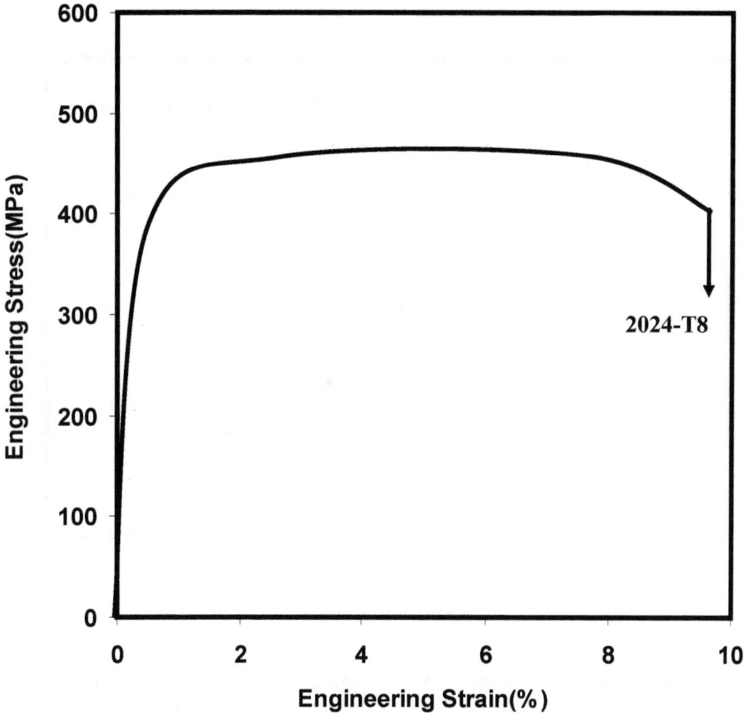

Figure 2. Variation of engineering stress with engineering strain for aluminum alloy 2024-T8

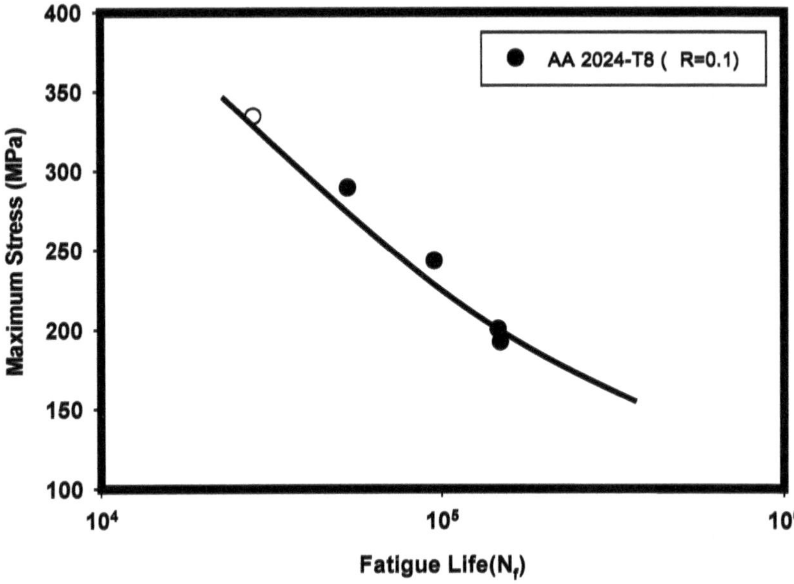

Figure 3. **Variation of maximum stress ($\sigma_{maximum}$) with fatigue life (N_f) for aluminum alloy 2024-T8 when deformed at room temperature and at load ratio of 0.1.**

reveals a trend of increasing fatigue life with decreasing maximum stress, a behavior shown by (i) other aluminum alloys belonging to the 2XXX series, (ii) most non-ferrous metals, and (iii) their composite counterparts [14-18]. However, the observed trend of increasing fatigue life with decrease in maximum stress is different from those shown by few other face-centered cubic metals that exhibit a well-defined endurance limit [8, 15, 18].

To understand and establish the influence of alloy plate ductility on high cycle fatigue response, quantified by cyclic fatigue life, the maximum stress-fatigue life test data is re-plotted in terms of elastic strain amplitude in a manner suggested and recommended by Hassen and co-workers [19], where the maximum elastic strain ($\sigma_{maximum}$ / E) is the maximum stress normalized by elastic modulus of the specimen in the specific orientation of the alloy plate. Variation of maximum elastic strain with fatigue life is shown in **Figure 4** and reveals a linear trend. This is done to essentially account for the influence of stiffness of the aluminum alloy as a function of test specimen orientation. In an attempt to rationalize the intrinsic influence of microstructure of the alloy plate on high cycle fatigue response, the maximum stress is normalized with respect to ultimate tensile strength (UTS) and its variation with fatigue life quantified in terms of cycles to failure (N_f) as shown in **Figure 5**

A comparison of the maximum stress versus fatigue life response of this alloy in the T8 temper with alloy 2024 in the T3 temper and alloy 2524 in the T3 temper at a constant load ratio of 0.1 is shown in **Figure 6**. This figure reveals the alloy in the T8 temper to have noticeably inferior fatigue resistance when compared to the same alloy in the T3 temper and the commercial aluminum alloy 2524 at all values of maximum stress. At equivalent values of maximum stress, and identical load ratio of 0.1, the degradation in cyclic fatigue life experienced by the T8 temper is as high as 300 percent over the entire range of stress when compared with the T3 temper.

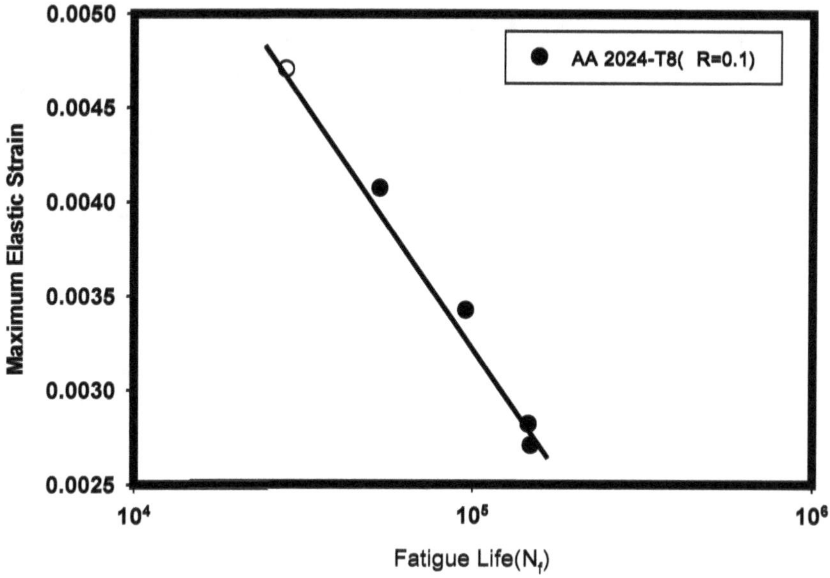

Figure 4. **Variation of maximum elastic strain ($\sigma_{maximum}$ / E) with fatigue life (N_f) for aluminum alloy 2024-T8 cyclically deformed at load ratio of 0.1.**

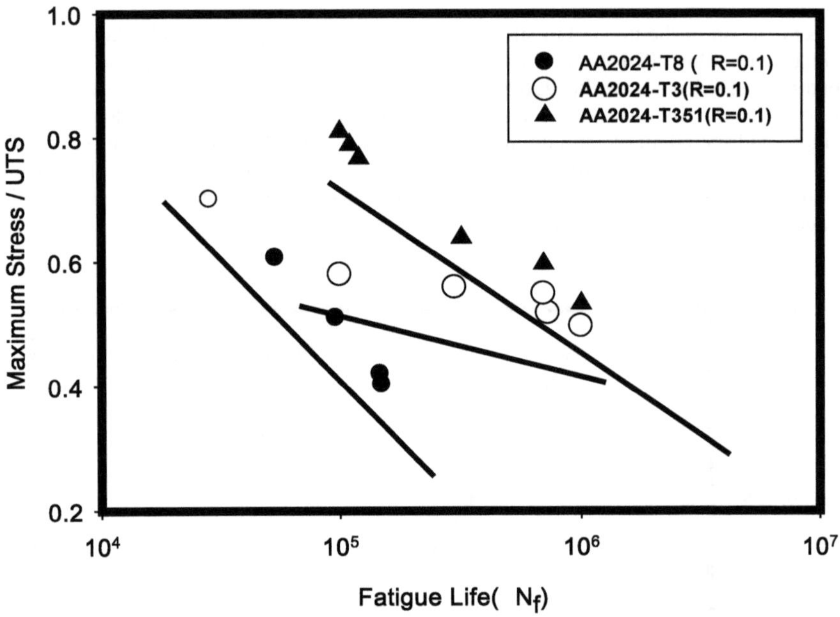

Figure 5. **The variation of ratio of maximum stress / tensile strength [$\sigma_{maximum}/\sigma_{UTS}$] with fatigue life ($N_f$) for aluminum alloy 2024-T8.**

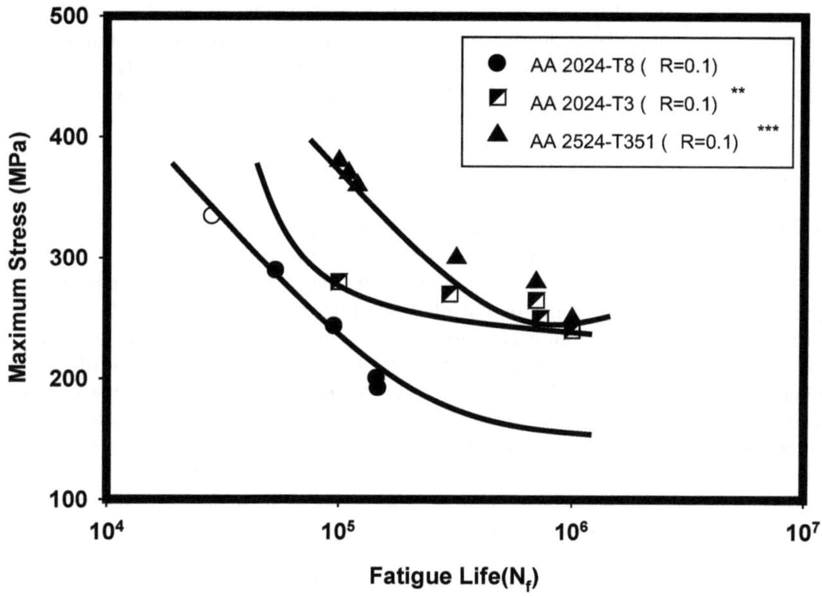

Figure 6. **A comparison of maximum stress versus fatigue life response of aluminum alloy 2024 in the T8 temper with the T3 temper and alloy 2524 in the T351 temper when cyclically deformed at a load ratio of 0.1.**
**** from Reference 15; *** from Reference 19**

High Cycle Fatigue Fracture

Examination of the fracture surfaces of the cyclically deformed and failed test specimens was done in a JEOL scanning electron microscope (SEM) at:

(i) Low magnifications to identify the regions specific to fatigue crack initiation and final fracture (overload).

(ii) At higher allowable magnifications in the fatigue and overload regions to identify the regions of microscopic crack formation, early microscopic growth and other fine scale features on the fracture surface.

Representative fracture features of the deformed and failed test specimens taken from: (i) high maximum stress resultant short fatigue life, and (ii) low maximum stress and resultant enhanced fatigue life, are shown in **Figures 7-9.**

At the load ratio of 0.1 and a maximum stress of 334 MPa the macroscopic fracture mode was essentially normal to the far field stress axis (**Figure 7-a**). High magnification observation of the cyclic fatigue fracture surface in the region of early microscopic crack growth revealed microscopic voids of varying size and shape along with fine microscopic cracks **(Figure 7-b).** In the region immediately prior to the onset of unstable crack growth and overload the fracture surface revealed a population of dimples and macroscopic voids intermingled with fine microscopic voids. These features are clearly indicative of the occurrence of locally ductile failure mechanisms [**Figure 7-c**]. Well into the region of unstable crack growth the fracture surface revealed the coalescence of the fine microscopic cracks to form a macroscopic crack prior to failure by separation (**Figure 7-d**).

Sample of this 2024-T8 alloy that was cyclically deformed at a lower value of maximum stress, i.e., 243 MPa with a resultant fatigue life of 95,877 cycles, the overall fracture surface morphology was essentially normal to the far-field stress axis **(Figure 8-a**). High magnification observation of the region of stable microscopic crack growth revealed a combination of voids of varying size, dimples, few and far between fine microscopic cracks (**Figure 8-b**). These features are indicative of the ductile and brittle failure mechanisms occurring at the local level. The region of early unstable crack growth revealed cracked second-phase particles intermingled with fine microscopic voids (**Figure 8-c**). These features are reminiscent of the occurrence of both brittle and ductile failure mechanisms at the fine microscopic level. The region of unstable crack growth was distinct and revealed an observable population of shallow dimples adjacent to the propagating macroscopic cracks (**Figure 8-d).**

Sample of the alloy that was cyclically deformed at a lower maximum stress of 200 MPa and resultant enhanced fatigue life of 146.684 cycles the transgranular fracture regions of the sample surface revealed voids, dimples and intermingled with microscopic cracks (**Figure 9-a**). Careful high magnification observation of the transgranular region revealed the dimples to be of varying size and shape (**Figure 9-b).** Approaching the region of unstable crack growth and immediately prior to failure the fracture surface revealed a combination of microscopic voids and cracked particles (**Figure 9-c**); features indicative of the occurrence of both ductile and brittle failure mechanisms at the local level.

Mechanisms Governing Cyclic Fatigue Fracture

During cyclic deformation of this precipitation hardened aluminum alloy (i.e., 2024) in the T8 condition or temper there does exist a tendency for localized inhomogeneous deformation arising from repeated shearing or destruction of the matrix strengthening precipitates at the fine microscopic level [3, 21-23]. This gradually results in the localization of deformation and resultant accumulation if strain at the fine microscopic level. When the local concentration of stress as a direct consequence of localization of strain at the coarse second phase particles and grain boundaries, exceeds the stress that is required for crack nucleation, microscopic crack initiation is favored to occur [22-24]. The presence of an observable population of both coarse and intermediate size second-phase particles in the alloy microstructure coupled with a microstructure, i.e., the T8 temper, which favors localized inhomogeneous deformation, is conducive for the nucleation, growth and eventual coalescence of the fine microscopic void to occur at the applied level of maximum cyclic stress.

The initiation of the fine microscopic cracks in the alloy microstructure and their concurrent growth during repeated cyclic loading can be ascribed to the progressive accumulation of micro-plastic deformation at the localized level. The accumulation of micro-plastic damage is exacerbated by stress-microstructure interactions that occur during each cycle of loading. It is this irreversible plastic deformation that is responsible for the initiation and resultant growth of both the fine microscopic cracks and macroscopic cracks through the alloy microstructure. The presence of voids, of varying size and shape, transforms the polycrystalline aluminum alloy to a composite with the voids being safely considered as particles having zero stiffness [24-26]. The initiation of the fine microscopic voids at the coarse second-phase and intermediate size particles in the alloy microstructure and their incremental growth during repeated cyclic deformation culminating in their coalescence does enable in providing small increments of crack tip extension. The finer voids are favored to form at the intermediate-size and smaller size second-phase particles dispersed through the alloy microstructure. Since the voids are intrinsically softer than the hardened grains in the alloy matrix, the local strain is significantly elevated for the microscopic voids causing as a result an increase in their volume fraction during continued cyclic loading.

Table 1: Nominal chemical composition of Aluminum Alloy 2024 (In weight percent)

Cu	Mg	Mn	Fe	Si	Zn	Ti	Aluminum
4.45	1.36	0.71	0.5	0.5	0.2	0.15	Balance

Table 2: Room temperature tensile properties of aluminum alloy 2024

Material	Temper	Elastic Modulus (GPa)	Yield Strength (MPa)	UTS (MPa)	Elongation GL=0.5" (%)
AA 2024	T8	71	445	476	8.84
	T3	73.1	345	483	18
	T4	73.1	324	469	19

Conclusions

Based on a study of the high cycle fatigue and final fracture behavior of aluminum alloy 2024 in the T8 temper, following are the key findings:

1. The as-provided alloy plate revealed large recrystallized grains of non-uniform size and shape with a non-uniform distribution of the coarse second phase particles. At frequent intervals particle agglomeration or clustering was evident.

2. The yield strength of the alloy plate in the T8 temper is noticeably higher than the yield strengths in the T3 and T4 temper. The ultimate tensile strength is about similar in the T8, T3 and T4. In the T8 temper the ultimate tensile strength is only marginally higher than the yield strength indicating low strain hardening capability of the alloy beyond yield.

3. The maximum stress versus fatigue life (N_f) curve reveals a trend that is shown by most non-ferrous metals and their composite counterparts. However, the observed trend is different from those shown by other face-centered cubic metals that exhibit a well-defined endurance limit.

4. At equivalent values of maximum stress the fatigue life of the alloy sample in the T8 temper is noticeably inferior to the T3 temper and the commercially aluminum alloy 2524 in the widely used T3 temper.

5. For this as-provided alloy plate fatigue fracture surface revealed noticeably different topographies at the different values of maximum stress. On a fine microscopic scale the nature, morphology and volume fraction of the intrinsic features on the fatigue fracture surface were found to vary with maximum stress and resultant fatigue life.

6. The presence of a noticeable population of voids of varying size, dimples of varying size and shape and intermingled with microscopic and macroscopic cracking along the grain boundaries is indicative of the occurrence of both ductile and brittle failure mechanisms at the fine microscopic level.

Acknowledgements:

The authors (TSS, SV and MK) extend sincere thanks, gratitude and appreciation to The **BOEING Commercial Aerospace Company** (***Phantom Division***, St Louis, MO, USA) for providing the material used in this study and for funding this research exercise.

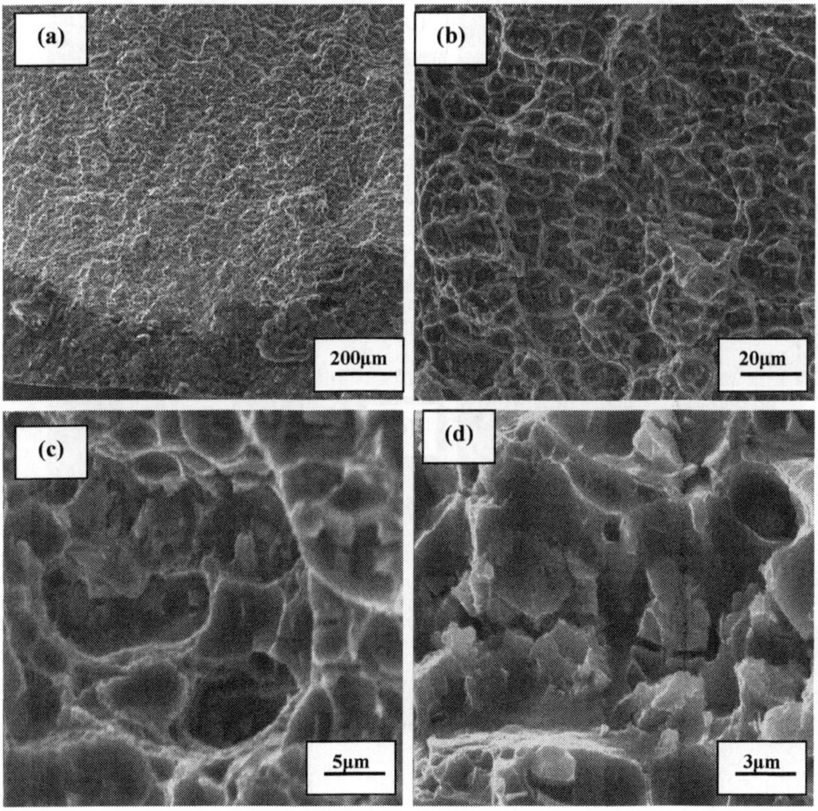

Figure 7. Scanning electron micrographs of the high cycle fatigue fracture surface of aluminum alloy 2024-T8 deformed at maximum cyclic stress of 334 MPa (R = 0.1), fatigue life = 28,344 cycles, showing

(a) Overall morphology
(b) High magnification of (a) showing microscopic voids of varying size and shape along with fine microscopic cracks.
(c) Dimples, macroscopic voids and microscopic voids in the region of unstable crack growth
(d) Microscopic cracks coalesce to form a macroscopic crack in the region of unstable crack growth.

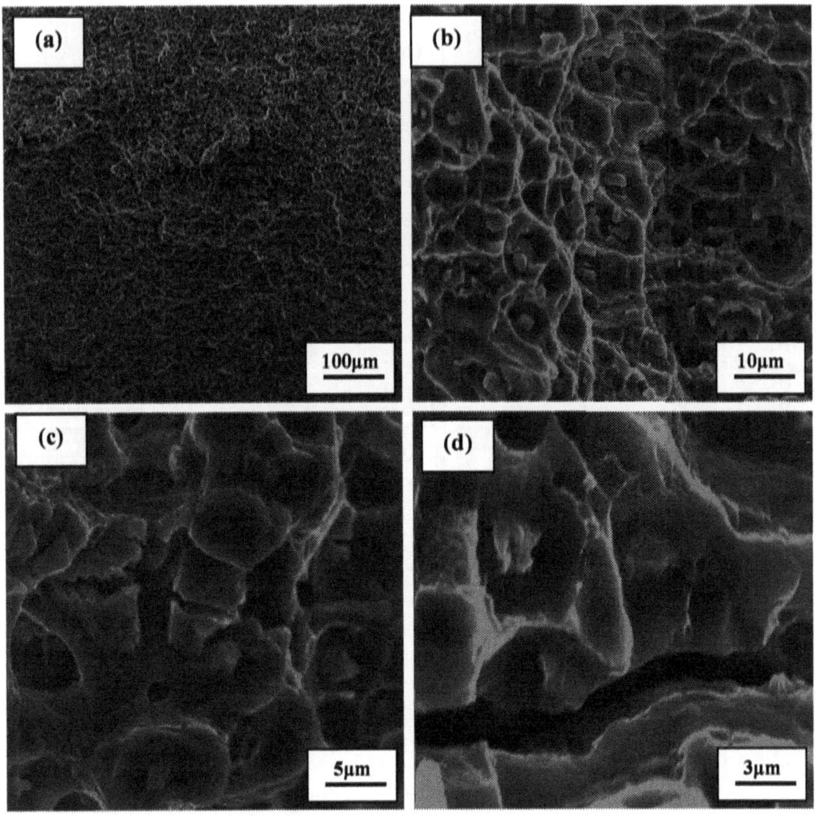

Figure 8. Scanning electron micrographs of the high cycle fatigue fracture surface of aluminum alloy 2024-T8 deformed at maximum cyclic stress of 243 MPa (R = 0.1), fatigue life = 95,877 cycles, showing

(a) Overall morphology
(b) High magnification of (a) showing voids of varying size, dimples and microscopic cracks.
(c) Cracked particles and microscopic voids in the region of unstable crack growth
(d) Shallow dimples adjacent to macroscopic crack in the region of unstable crack growth.

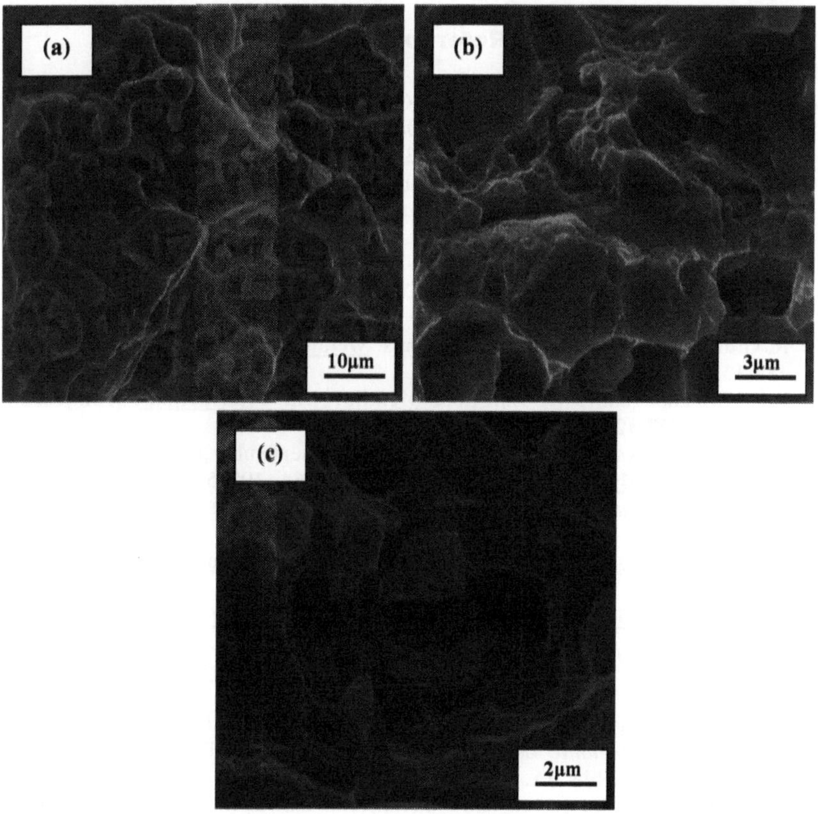

Figure 9. **Scanning electron micrographs of the high cycle fatigue fracture surface of aluminum alloy 2024-T8 deformed at maximum cyclic stress of 200 MPa (R = 0.1), fatigue life = 146,684 cycles, showing**

(a) Voids, dimples and microscopic cracks on the transgranular fracture surface

(b) High magnification of (a) showing dimples of varying shape and size.

(c) Microscopic voids and cracked particle in the region of unstable crack growth.

References

1. E.A. Starke, Jr., in *Aluminum Alloys: Contemporary Research and Applications* (editors: A. K. Vasudevan and R. D. Doherty), Treatise in Materials Science and Technology, Academic Press, Volume 31, 1989, pp. 35-45.
2. A.K. Vasudevan, R.D. Doherty and S. Suresh: in *Aluminum Alloys: Contemporary Research and Applications* (edited by A.K. Vasudevan and R. D. Doherty), Treatise in Materials Science and Technology, Academic Press, Volume 31, 1989, pp. 446-456.
3. J.T. Staley: Aluminum Alloys and Composites, in *Encyclopedia of Physical Science,* Vol. 1, 1992, pp. 591-611.
4. R. Grimes, M. J. Stowell and B. M. Watts: **Metals Technology**, Vol. 3, 1976, pp. 154-155.
5. R.H. Bricknell and J.W. Edington: **Metallurgical Transactions**, Vol. 10Am, 1979, pp. 1257-1267.
6. Aluminum Alloys in the 1970s: **Metal Progress**, September 1970, pp. 68-79.
7. M. O. Speidel: **Metallurgical Transactions**, Vol. 6A, 19875, pp. 631-640.
8. S. Suresh: *Fatigue of Materials*, Cambridge University Press, Cambridge United Kingdom, 1991.
9. H.J. Miller: **Fatigue of Engineering Materials and Structures**, Vol. 5, 1982, pp. 223-232.
10. J. Lankford: **Fatigue of Engineering Materials and Structures**, Vol. 5, 1982, pp. 233-243.
11. M. Sauzay and T. Jourdan**:** **International Journal of Fracture**, Vol. 141, 2006, pp. 431-466.
12. American Society for Testing and Materials: Standard E-8-10: "*Conducting Tension Test on Metallic Materials*," American Society for Testing and Materials (ASTM), Philadelphia, PA, USA, (2010).
13. American Society for Testing and Materials: Standard E-466-02: "*Conducting Constant Amplitude Axial Fatigue Tests on Metallic Materials*," American Society for Testing and Materials (ASTM), Philadelphia, PA, USA, (2002).
14. T. S. Srivatsan and M. Al-Hajri: **Composites**, Part B, Viol. 33, 20023, pp. 391-404.
15. T.S. Srivatsan, S. Vasudevan, Lisa Park and R. J. Lederich: **Materials & Design**, 2007,
16.
17. U. Alfaro, T. Ghidini and C. Dalle Donne**:** **"**Fatigue of pre-corroded 2024-T3 Friction Stir Welds' Experiment and Prediction",.
18. Fatigue and Fracture Handbook, ASM Metals Handbook Volume 19, American Society for Materials International, Materials Park, Ohio, 2009, pp.
19. D.F. Hassen, C.R. Crowe, J.S. Ahearn and D.C. Cooke: *Failure Mechanisms in High Performance Materials* (edited by J.E. Early, T.R. Shives and J.H. Smith), The Minerals, Metals and Materials Society, TMS, Warrendale, PA, USA, 1984, pp. 147-157

20. T.S. Srivatsan, Nurudeen Balogan and T. Quick in *Fatigue of Materials: Advances and Emergences in Understanding* (edited: T.S. Srivatsan and M. Ashraf Imam) The Minerals, Metals and Materials Society, TMS, Warrendale, PA, SA, 2010, pp. 339-354
21. T.S. Srivatsan, S. Anand and P. Magnusen: ***Materials Science and Engineering***, Vol. A314, 2001, pp. 118-123.
22. T.S. Srivatsan, D. Lanning and K. K. Soni: **Journal of Materials Science**, Vol. 28, 1993, pp. 3205-3211.
23. T.S. Srivatsan, S. Sriram, D. Veeraraghavan and V. K. Vasudevan: **Journal of Materials Science**, Vol. 32, 1977, pp. 2883-2889.
24. T.S. Srivatsan and Satish Vasudevan " The Science, Technology and Applications of Aluminum and Aluminum Alloys," *Advanced Structural Materials* (editors: Winston O Soboyejo and T.S. Srivatsan), CRC Press, Taylor & Francis Group, pp. 225-274.
25. R.H. Van Stone and J.A. Psioda: **Metallurgical Transactions**, Vol. 6A, 1975, pp. 672-677.
26. R.H. Van Stone, T.B. Cox, J.R. Low and J.A. Psioda: **International Materials Reviews,** Vol. 30, 1985, pp. 1567-177.

Fatigue of Materials II Advances and Emergences in Understanding
Edited by: T.S. Srivatsan, M. Ashraf Imam, and R. Srinivasan
TMS (The Minerals, Metals & Materials Society), 2013

FRACTOGRAPHIC OBSERVATIONS ON THE MECHANISM OF FATIGUE CRACK GROWTH IN ALUMINIUM ALLOYS

R.C. Alderliesten[1], J. Schijve[1], M. Krkoska[2]

[1] Structural Integrity & Composites, Delft University of Technology, Kluyverweg 1, 2629 HS, Delft, **The Netherlands**
[2] Material Innovation Institute, Mekelweg 2, 2628 CD Delft, **The Netherlands**

Keywords: Fatigue crack propagation, Fractography, Aluminum

Abstract

Special load histories are adopted to obtain information about the behavior of the moving crack tip during the increasing and decreasing part of a load cycle. It is associated with the crack opening and closure of the crack tip. Secondly, modern SEM techniques are applied for observations on the morphology of the fractures surfaces of a fatigue crack. Information about the cross section profiles of striations are obtained. Corresponding locations of the upper and the lower fracture surface are also explored in view of the crack extension mechanism. Most experiments are carried out on sheet specimens of aluminum alloys 2024-T3, but 7050-T7451 specimens are also tested in view of a different ductility of the two alloys.

Introduction

The mechanism of fatigue crack growth in aluminum alloys has been studied in many investigations in the past. A major observation was the occurrence of striations indicating a cycle-by-cycle crack growth phenomenon. In addition, several models were presented for the cyclic behavior of the crack tip profile on a microscopic level. Two well known models were proposed by McMillan and Pelloux [1] and by Laird [2] respectively. Differences between the two models are associated with the geometry of the crack tip, as illustrated in Figure 1. In the first model, the crack tip is supposed to be sharp during increasing stress and to remain sharp during decreasing stress until the crack is closed. The closure starts at the tip of the crack.
In the Laird model the tip is supposed to be rounded and to remain rounded during decreasing stress. It has been thought that the first model is the more likely one for brittle materials, whereas the second model may be applicable to highly ductile materials. Experimental validation of this hypothesis is difficult and so far it has not converged to a generally accepted understanding.

Another important historical observation was the detection of plasticity induced crack closure by Elber [3]. It implies that during unloading the crack is closed while the specimen is still under tension. Closing starts at the crack tip and continues behind the crack tip while the applied stress can still be positive. This type of crack closure on a macroscopic scale has led to various proposals for the prediction of fatigue crack growth by employing fracture mechanics.

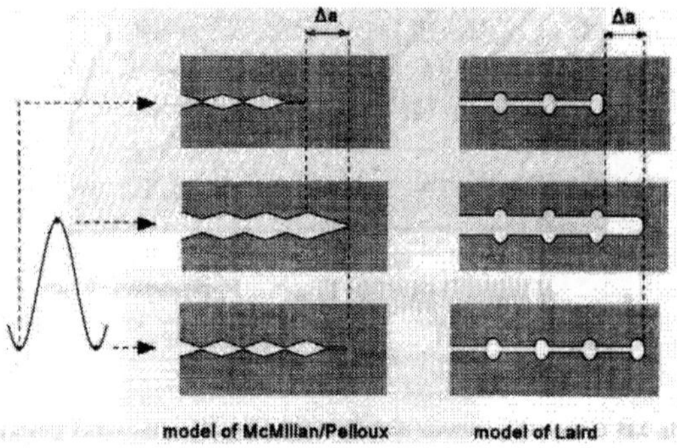

Figure 1. Two models for striation formation during fatigue crack growth [1,2]

The intention of the present research is to shed more light on the microscopic mechanism of the crack extension in a load cycle during both intervals of the load cycle, i.e. during increasing stress and decreasing stress. Special techniques were developed to achieve information about the micro-plastic deformation during crack tip opening and crack tip closure. Moreover, a new morphology technique was used to explore the fatigue fracture surfaces. It was thought to be essential to examine the fracture surfaces at exactly corresponding locations of the upper surface and the lower surface of a fatigue crack.
The experiments and specimens are described first in the next section, followed by the applied microscopy techniques. The results presented include several selected pictures obtained with scanning electron microscopy (SEM). The results are discussed followed by some summarizing conclusion in the last section.

Table 1 Mechanical properties of the tested aluminum alloys

	$S_{0.2}$ (MPa)	S_U (MPa)	$S_{0.2}$ / S_U	δ (%)
2024-T3	337	481	0.70	20
7050-T7451	441	510	0.86	9

Experiments

The specimen is illustrated in Figure 2. It is a slightly tailored sheet specimen with a fairly sharp (electro discharge machined) edge notch to initiate a fatigue crack. Two sheet materials were used, 2024-T3 and 7050-T7451. The mechanical properties of these alloys are given in Table 1. It is well known that sheet material of the 7000 series alloys have a lower ductility than the sheet material of the 2000 series alloys. As a result, fatigue cracks are growing faster in the 7000 series material. It may imply that the fatigue crack growth mechanism is not similar for the two alloy series. For that reason some tests were also carried out on sheet specimens of the 7050-T7451 sheet material.

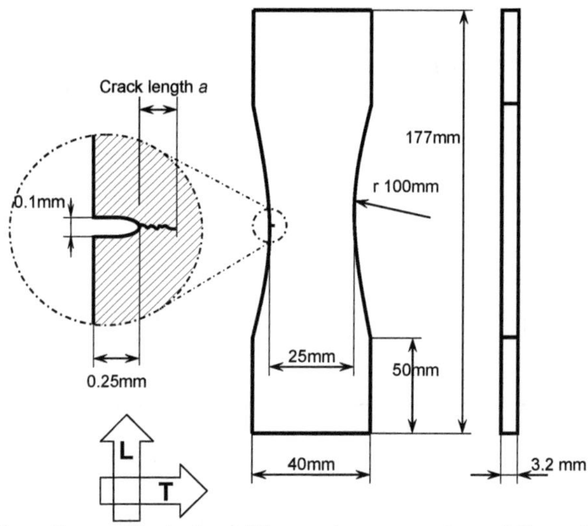

Figure 2. Illustration of the specimen geometry and dimensions

All tests were performed using a 100 kN MTS servo hydraulic machine with a frequency of 10 Hz in an ambient environment (air, temperature). In view of large negative loads anti-buckling plates were attached to the specimens. An opening was machined in the plates to allow optical measurements of the crack propagation with a travelling microscope.
Several load sequences were adopted in order to obtain specific information about the crack growth mechanism. Examples of some typical load histories applied, are shown in Figures 3 and Figure 5. The load history in Figure 3 (a) consists of blocks with 100 baseline cycles at R=0.5 interspersed with underloads (UL) of different magnitudes. The load history given in Figure 3 (b) contains UL of R=0 with in-between subsequent UL an increasing block of baseline cycles at R=0.5. Similar load histories as shown in Figure 3 were carried out with different numbers of cycles in the baseline blocs and also with more UL cycles interspersed between the blocks. Typical crack growth curves obtained are illustrated in Figure 4. The main purpose was to obtain new insights about the morphology of the fatigue fracture surfaces with modern electron microscope techniques.

The load histories in Figure 5 were used to obtain observations on the crack tip deformation occurring in a single cycle. The test was interrupted at a certain load level and then continued with a large number of cycles containing an amplitude of 7.5 MPa until specimen failure. It was expected that these small cycles would not change the profile of the crack tip. It can be considered as freezing the shape of the crack tip, which then can be observed with SEM. This freezing procedure was repeated at different stress levels during the raising part and the decreasing part of a load cycle in order to obtain a sequential impression of the micro plastic deformations occurring in one load cycle which ultimately creates a striation.

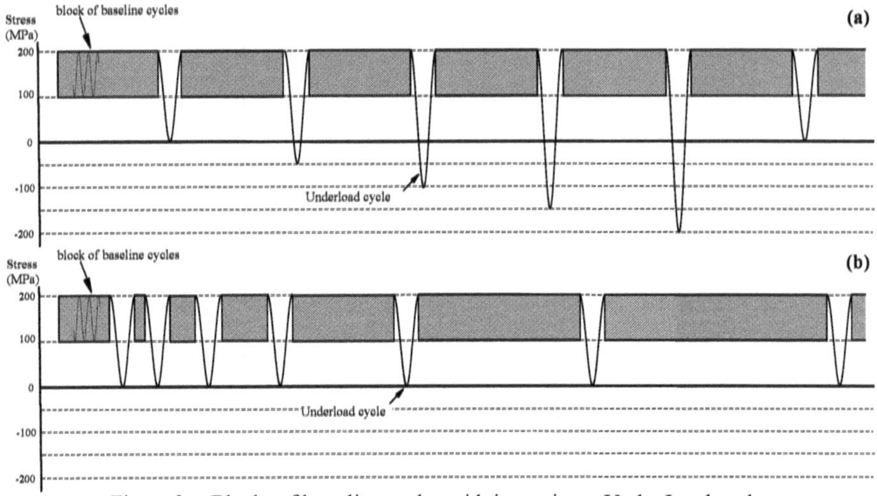

Figure 3. Blocks of base line cycles with intermittent Under Load cycles

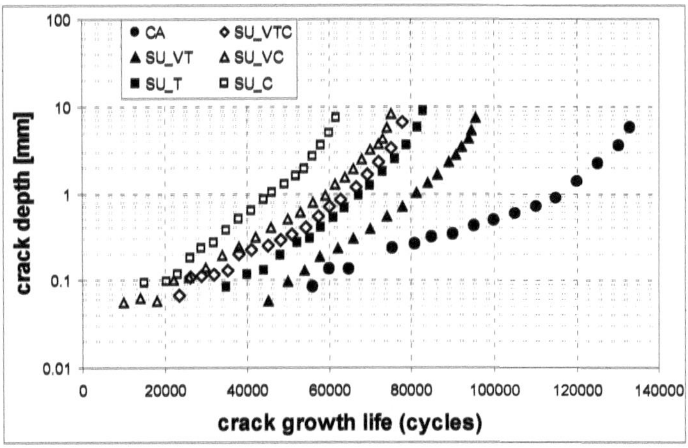

Figure 4. Crack growth curves for load sequences employing single tensile (▲■) and compressive (Δ □) underloads, compared with CA baseline cycles (•).

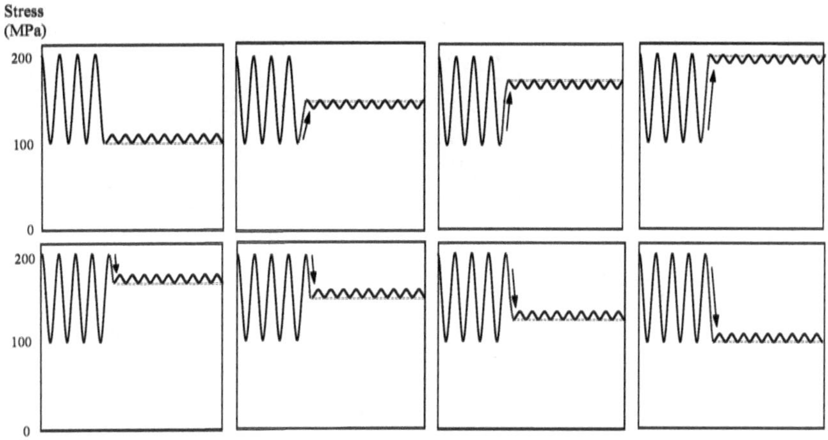

Figure 5 Load sequences adopted for observing changes of the crack tip geometry in the SEM. The baseline cycles were interrupted at different stress levels after which a large number of very small cycles was applied to extent the crack without disturbing the crack tip profile from the last load excursion.

Microscopy Techniques

A JOEL JSM7500F scanning electron microscope with secondary electron detector was employed. Useful images covered a range of crack length values from $a = 0.5$ mm to about $a = 6$ mm. Various SEM pictures were made in view of obtaining information about striation spacings as well as the cross sectional profile of striations. Attention was also paid to the variation of topographic observations along a crack front, i.e. variations in the thickness direction, to avoid generalization of so-called typical pictures.

Fractography on fatigue fracture surfaces traditionally utilizes two-dimensional projections of fracture surfaces. However, the third dimension is considered to be essential for describing striation deformations at the moment of the crack tip extension.
The three-dimensional characterization was performed in two successive steps. First, the two mating fracture surfaces were inserted in the scanning electron microscope. Observing mating locations on the upper and lower fracture surface was a difficult task, but necessary to assess deviations from mirror profiles.
Images were taken from the fracture surface first at 350 times magnification conditions, 5 kV accelerating voltage and working distance of 7 mm in the lower magnification mode in order to determine the general fracture surface orientation. These were digitally "stitched" together creating the single micrograph of the fracture surface that was created under cyclic load, see Figure 6.

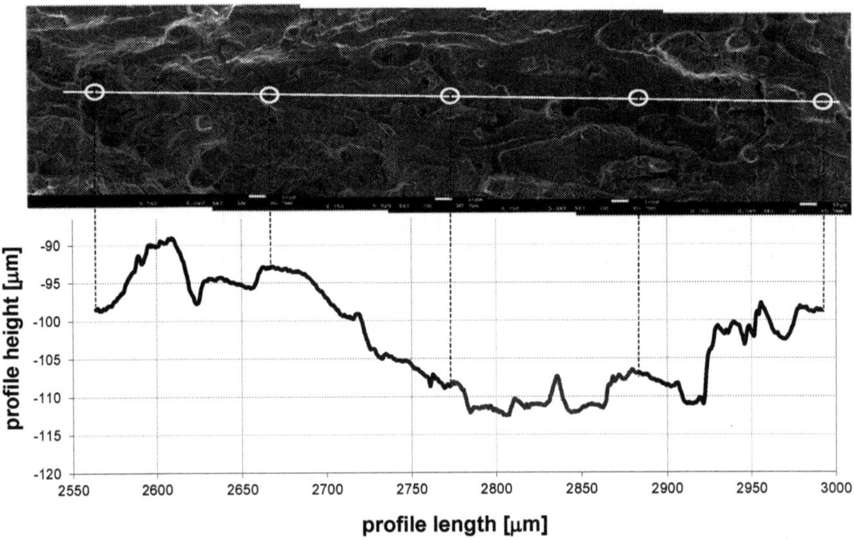

Figure 6. Example of micrograph's "stitching" with profiles obtained from pre-selected paths

Multiple crack paths were digitally drawn on the fracture surface from crack initiation up to the final failure area. An example of such path is illustrated in Figure 6. These paths were then used to obtain fracture surface profiles from individual digital elevations. The obtained profiles were plotted together creating the single fracture surface profile data set, like in Figure 6. The angle of fracture surface tilt was then determined by linear regression method for each fracture surface. A preliminary study in which magnifications of 50, 100, 200, 350 and 750 were investigated, illustrated that the results following this approach, converged for magnifications of 350 times and higher (i.e. fracture angle tilt no longer depended on the magnification level).

A special technique was developed to arrive at three-dimensional information of the morphology of the surface of fatigue cracks. The microscopic images were not only taken parallel to the electron beam, i.e. 0°, but also at tilting angles of -7° and 7°. These images were processed subsequently with the MeX program employing a TriCreator option for producing digital elevations.
Maintaining the same working distance and specimen tilting angles, micrographs were taken along the crack tip at higher magnification levels. The corresponding locations on both the upper and lower fracture surface were identified for each determined fracture profile. Subsequently, matching digital paths were selected on both fracture surfaces to enable plotting the crack tip opening geometry.

Fractographic Results

The first observation when comparing the fracture surfaces of 2024-T3 and 7050-T7451 is the difference in appearance. The fracture surface of 7050-T7451 reveals fracture features typical for brittle type fatigue fractures, whereas the 2024-T3 fracture is less flat and shows more ductile fracture features. This is illustrated in Figure 7.

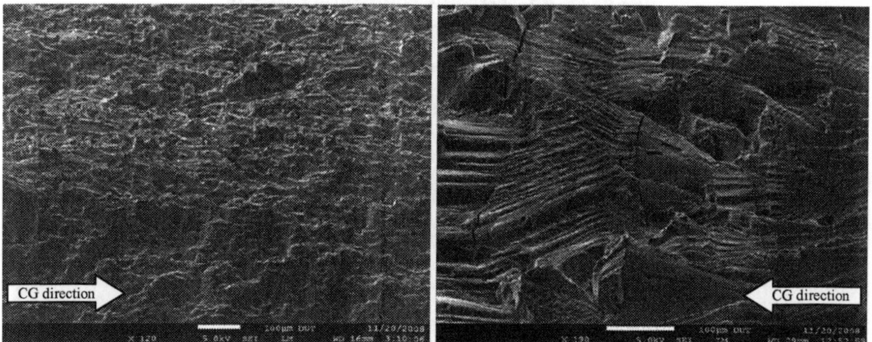

Figure 7. Comparison between the microscopic appearance of fatigue fracture surfaces created in 2024-T3 (left) and 7050-T7451 (right); arrows indicate crack growth direction.

It was observed that that crack initiation started immediately due to the sharp crack starter notch. However, initially the striation-like features indicated a crack growth rate which slowed down a little which is associated with building up a plastic wake of the crack. Near the notch, a fatigue fracture surface was formed with characteristic feather-like surface patterns related to the simultaneous growth of multiple crack nuclei. The coalescence of crack nuclei creates this typical fracture, illustrated in Figure 8. With subsequent growth of a single crack front these fracture features disappear and beyond approximately 1 mm crack length from the notch the fracture surface consisted of areas with relatively smooth but wavy appearance up to about 2.5 mm. At a larger crack length the roughness and irregularity increased, but still with clear and evident striation markings on the fracture surface.

Another observation is that the striation profiles varied along the crack length, and sometimes exhibited flattened striations indicating contact between upper and lower fracture surface. Results of the measurements about the striation profiles during the blocks of base line cycles are illustrated in Figure 9. The onset, i.e. the very beginning of the crack extension in a load cycle, was usually well defined as a straight part of the striation profile. However, after the maximum load was attained, the load reduction causes a significant crushing of the crack tip. The crushing operation has led to various patterns of the striations. Whereas simple roof-top striation shapes in image A1 in Figure 9 still look rather similar to the profile of the two-dimensional model of MacMillan, the general trend is that highly different profiles occur both along the crack length and along the entire crack front through the material thickness. Most striations observed contain features at their leading edge that indicate stretching of the material, while their irregular trailing edges indicate extensive reversed plastic deformation.

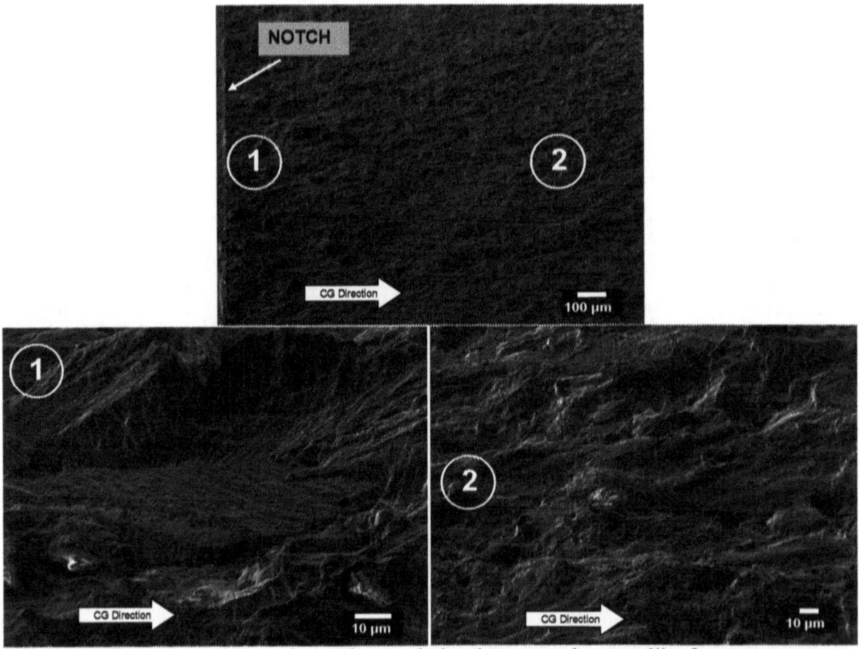

Figure 8. Illustration of the early crack growth development: cleavage-like fracture at very short crack lengths (1) and typical fatigue fracture at longer lengths (2) [5].

The fracture surface shown in Figure 10 was obtained with the load history shown in Figure 3 (b). The large striations are associated with the UL cycles, and in fact with the second load-increasing part of these cycles. The increasing crack growth increment between the UL cycles corresponds with the increasing number of cycles in the blocks of base-line cycles as indicated in Figure 3. Perhaps more important, the bright image of the UL striations suggest a crack extension in a direction deviating from the global crack growth direction during the blocks of base-line cycles. This crack plane orientation has an evident influence on the ridge formation at the moment of the UL application. At the same time, Figure 10 illustrates that the fracture surface is a rather rough surface, and not at all a flat surface on a microscopic level.

The latter observation is further supported with the fracture surface illustrated in Figure 11. Again the UL striations are clearly visible, but seem to become different in appearance once the local crack plane orientation diverts from the general crack plane. In addition it presents a typical image of the variation of striation patterns along a crack front with disturbing areas of the striation patterns.

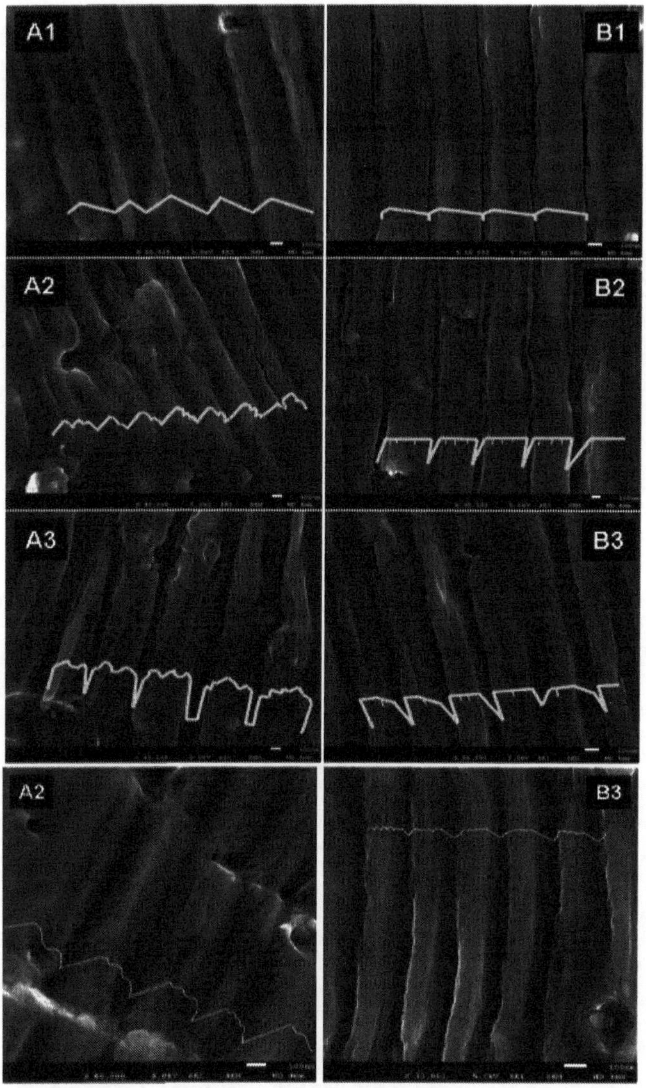

Figure 9. Different striation profiles; well developed (left) and flattened (right). Clearly visible is the difference in appearance between leading and trailing edge. The white magnification bar corresponds to 0.1 μm; the magnification is quite large.

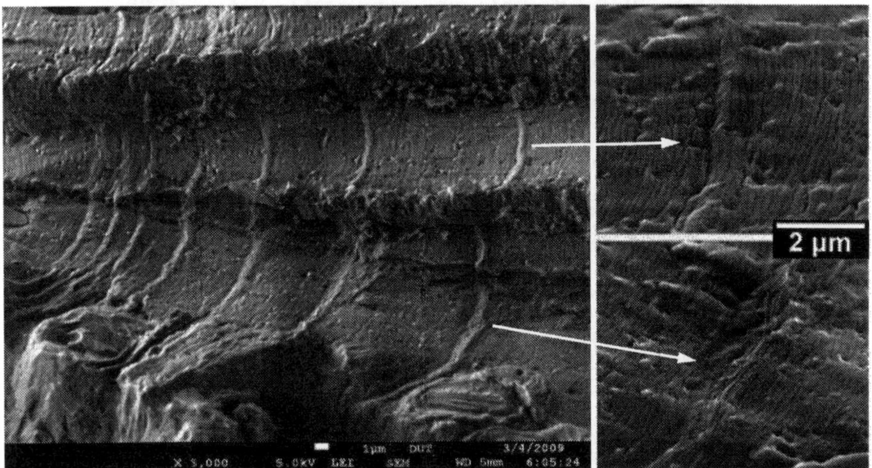

Figure 10. Fracture surface created with the load sequence illustrated in Figure 3 (b) highlighting the influence of crack plane orientation on the ridge shapes (K_{max} = 14 MPa√m). Crack propagates from left to the right [4].

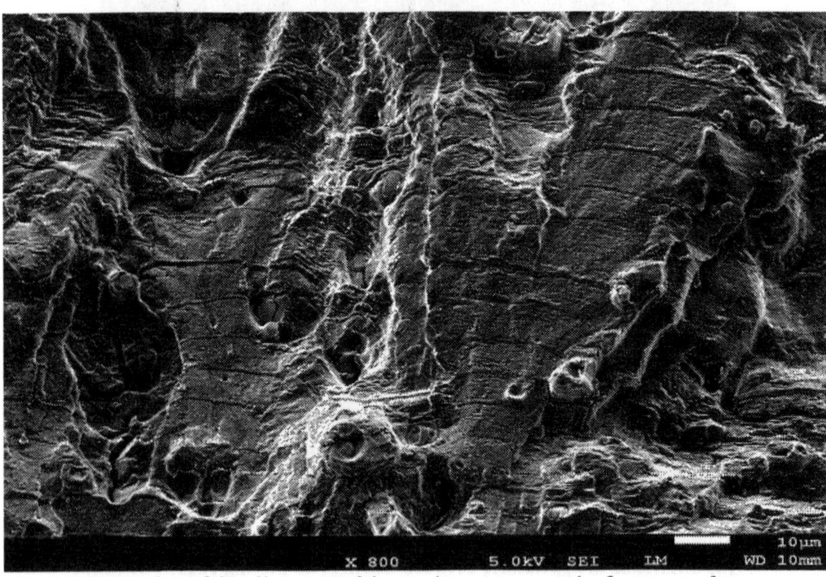

Figure 11. Illustration of the alignment of the crack contour over the fracture surface

Discussion

The assumption that the upper fracture surface is a mirror image of the lower fracture surface was observed to be incorrect. Although flat and smooth at macroscopic level, the topography of the fracture surfaces at microscopic level is rather distinct. This implies that locally the fracture surface may be angled with respect to the dominant fracture orientation leaving features at one side of the fracture surface (like for example fissures or ridges) that were not visible at the other surface at that particular location.

In addition, the crack opening profile was in most cases observed not to be a mirror image and very dependent on the location along the crack front. This is in contradiction with for example the models of McMillan/Pelloux and Laird, illustrated in Figure 1, which ignore the third dimension. This was identified using the fracture surface profiling technique mentioned before where corresponding paths were selected on both the upper and lower fracture surface for which the profile was determined. Accounting for the fracture profile determined at lower magnification levels, see Figure 6, an accurate crack tip opening profile could be determined. A result for matching surfaces is illustrated in Figure 12, indicating various matching features that are not mirror images from each other.

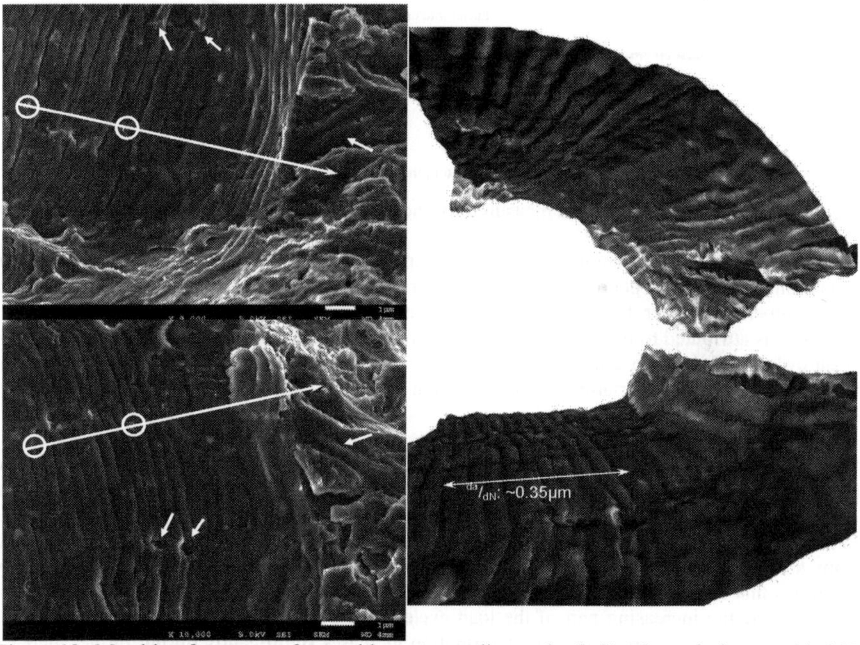

Figure 12. Matching fracture surfaces with corresponding paths (left). The technique enabled 3D representations of matching surfaces (right).

Determining the crack profile in conjunction with the general crack path orientation determined at magnification of 350, the crack tip opening profile can be reconstructed along the path indicated in Figure 12. The crack opening profile along that path is shown in Figure 13. Evident for the illustrated case is the non-symmetric appearance of the crack opening.

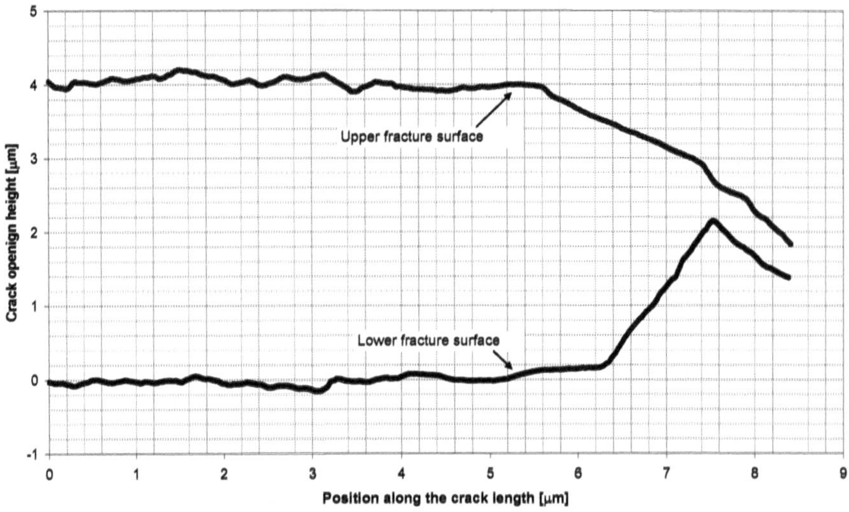

Figure 13. Crack opening profiles measured for the path highlighted in Figure 12.

This asymmetric crack opening near the crack tip evidently has consequences for the striation formation on both the upper and lower fracture surface. It was observed that at identical locations of the upper and lower fracture surface the striations did not match in size and shape. This mismatch is attributed to the asymmetric deformation of the crack tip opening, where subsequent closure during unloading collapsed a formed ridge on one fracture surface (lower surface in Figure 13) into a large striation, without leaving a similar feature on the other surface (upper surface in Figure 13).

This underlines the importance of the current study on the formation of striations during the increasing and decreasing parts of the load cycles. To illustrate the observed phenomena, Figure 14 depicts SEM images taken at the location on the fracture surface where the crack tip was formed during the last (part of the) base line cycle, before the situation was 'frozen' by applying many small amplitude cycles. The first case, shown in the upper left side of the figure, clearly illustrates that the amount of plastic deformation (part of striation formation) is relatively small until halfway the increasing part of the load cycles. The next to images (upper right and lower left), show that the amount of plastic deformation increases significantly during further increase of the load. Similarly, the first part of the decreasing part of the load cycle reverses most plasticity formed during the increasing part, collapsing the ridge back into a striation.

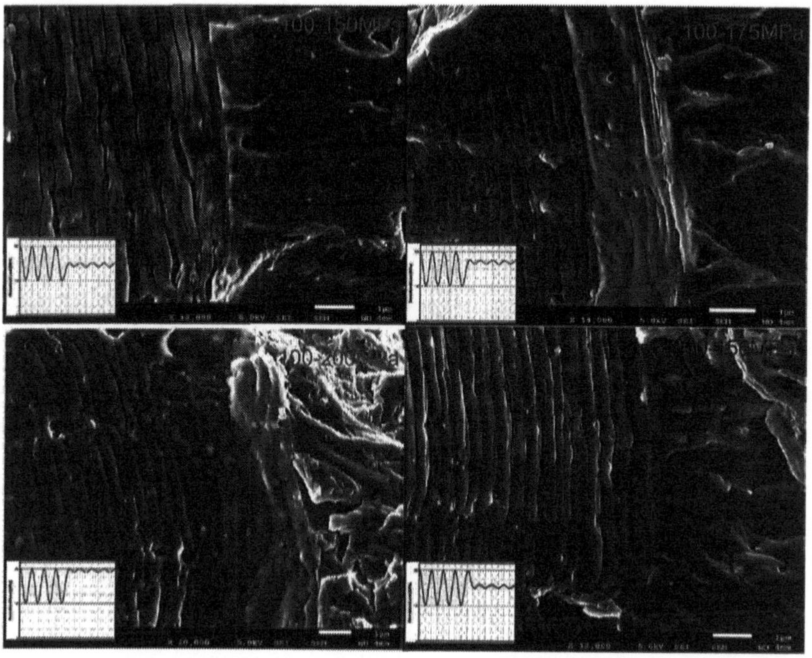

Figure 14. Illustration of the striation formation during the increasing (a-c) and decreasing part of the load cycle (d)

All previous pictures, comments and conclusions were associated with experiments and fractography of 2024-T3 specimens. Some research has been carried out with specimens of the 7050-T7451 alloy. Results will be published later. The two comparative pictures presented here in Figure 7 already suggest indeed a large difference between the behavior of the two alloys. It may well be associated with the different ductility of the alloys which may affect the process of forming striations.

Conclusions

The present investigation is concerned with the crack growth mechanism at the tip of a fatigue crack in the aluminum alloy 2024-T3. Special load histories and modern SEM techniques were employed. Some summarizing conclusions are:

1. The two-dimensional models of fatigue crack growth proposed in the past are a simplification of the three-dimensional crack growth behavior observed in the present research. It is a pronounced up and down phenomenon leading to a microscopically randomized topography.

2. The crack tip plasticity during the load increasing part of the cycle occurs largely near the maximum load of the cycle. It is then immediately followed by reversed plasticity still at the upper part of the load cycle.
3. Significant differences between the striations at the upper side of a fatigue crack and the lower side were observed. The profile of the striations at both sides should not be considered to be mirror images.
4. Micrographs along a crack front exhibit a significant diversity of fracture surface and striation profiles, and not a so-called typical picture.

References

[1] J.C. McMillan and R.M.N. Pelloux, "Fatigue crack propagation under program and random loading," ASTM STP 415, (1967), 505-535.

[2] C. Laird, "The Influence of Metallurgical Structure on the Mechanisms of Fatigue Crack Propagation", ASTM STP 415, (1967), 131-180.

[3] W. Elber, "The Significance of Fatigue Crack Closure," ASTM STP 486 (1971), p.230-242.

[4] M. Krkoška, S.A. Barter, R.C. Alderliesten, P. White, R. Benedictus, "Fatigue crack paths in AA2024-T3 when loaded with constant amplitude and simple underload spectra", Engineering Fracture Mechanics 77 (2010), 1857–1865.

[5] M. Krkoška, R.C. Alderliesten, R. Benedictus, "On the crack growth behaviour under simple compressive loads of 2024-T3 Aluminium alloy", proceedings of the 25th ICAF Symposium, Rotterdam, 27–29 May 2009.

Fatigue of Materials II Advances and Emergences in Understanding
Edited by: T.S. Srivatsan, M. Ashraf Imam, and R. Srinivasan
TMS (The Minerals, Metals & Materials Society), 2013

FATIGUE PREDICTIONS OF VARIOUS JOINTS OF MAGNESIUM ALLOYS

Kang, H.[1], Kari, K.[1], Getti, A.[1], Khosrovaneh, A.K.[2], Su, X.[3], Zhang, L.[4], Lee, Y.-L.[4]
[1]The University of Michigan-Dearborn, 4901 Evergreen Road, Dearborn, MI 48128
[2]General Motors Holdings, Warren, MI
[3]Ford Motor Corporation, Dearborn, MI
[4]Chrysler Corporation LLC, Auburn Hills, MI

Keywords: Fatigue, Self-Piercing Rivet Joint, Friction Stir Spot Weld, Linear Friction Stir Weld, AM60B

Abstract

In this project, a front shock tower of a passenger vehicle is developed with various magnesium alloys and joining methods. To predict the fatigue life of the joints in the structure, fatigue tests of various joint specimens including friction stir linear welding, self-piecing rivet joint with and without adhesive, and friction stir spot welding were conducted. The magnesium alloys used for the specimens are AM60 (cast), AM30 (extrusion), and AZ31 (sheet). Various finite element modeling techniques were attempted for simulating the various joints. Fatigue life prediction method for the joints was performed using the stress-life curve approach. The finite element modeling technique and the fatigue prediction method will be verified with fatigue tests of the actual front shock tower structure subjected to variable amplitude loadings in near future.

Introduction

Magnesium (Mg) alloys are applied to the various areas in automotive industry including powertrain, chassis, and body applications since the Mg alloys are 33% lighter than aluminum alloys and 75% lighter than steels and cast irons [1]. The joining of magnesium alloys since these components have to be connected to each other and/or to other materials in a vehicle structure. Various joining techniques were reported for magnesium alloys in application to the automotive industry [2-7]. Friction stir welding (FSW) is one of promising joining technique for magnesium alloys. FSW is a solid state joining process which was invented in 1991 [8]. A special designed tool creates frictional heat that softens a column of material underneath the tool while it rotates and transverses along the joint line. The softened material flows around the tool through extensive plastic deformation and is consolidated behind the tool to form a solid state continuous joint. Friction stir spot welding (FSSW) is a novel variant of the linear FSW process. FSSW is also used to join magnesium alloys [2, 4]. Self-piercing rivet (SPR) joint is getting more attentions because of the joining capability of dissimilar materials [9, 10].

The structural applications of the various joints for magnesium alloys will be the critical locations when the structure is subjected to service loads. Thus, it is necessary to understand the characteristics of the durability performance of those joints. Furthermore, fatigue life prediction methods for those joints are needed during the design phase of the structure. This paper discusses the attempts to develop fatigue prediction methods based on finite element analysis (FEA) results and fatigue test results of FSW, FSSW, SPR and SPR with adhesive joints.

Joint Types and Configurations of the Magnesium Alloys

Four types of magnesium alloy joints were constructed to understand the durability performance of the joints, which include friction stir spot weld (FSSW), friction stir weld (FSW), self-piercing rivet (SPR) joint, and SPR with adhesive joint. The magnesium alloys used to construct the joints are AZ31, AM30 and AM60. For all joint types, lap-shear specimens were constructed for fatigue testing. FSSW joint specimens were constructed with AZ31 to AZ31 alloy, and the specimen geometry and dimensions are as shown in Figure 1 [11]. The specimen geometry and dimensions for FSW are shown in Figure 2. The geometry and dimensions for SPR and SPR with adhesive joint specimens are shown in Figure 3. SPR, FSW, SPR, and SPR with adhesive joints had various combinations of magnesium alloys as shown in Table 1. Fatigue tests for FSW, SPR, and SPR with adhesive joints were conducted by AET Integration, Inc [12].

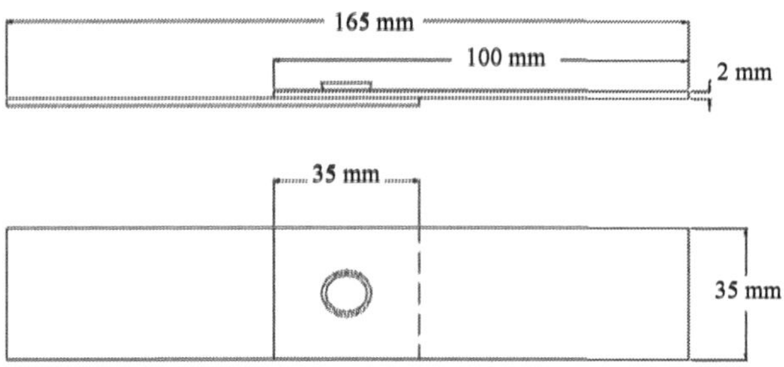

Figure 1. Geometry and dimension of FSSW lap-shear specimen for magnesium alloys [11].

Modeling of Various Joint Types

Finite element analysis (FEA) models for FSSW, FSW, SPR, and SPR with adhesive joints were developed with a commercial pre-processor called HyperMesh, and the linear elastic finite element analyses were conducted with a commercial FEA solver called Radioss. The base material was modeled with 2-dimensional (2-D) shell elements for sheet (AZ31) and extrusion (AM30) parts. However, the cast magnesium alloy, AM60, was modeled using 3-dimensional (3-D) tetrahedral elements since all cast components in vehicle structure are modeled with 3-D tetrahedral elements. The material properties used in this study are shown in Table 2. One of the most important factors considered for modeling these joints was whether these modeling techniques can be used in a full vehicle structure that consists of thousands of these joints. Thus, the modeling techniques for representing these joints should be simple enough but provide reasonable accuracy.

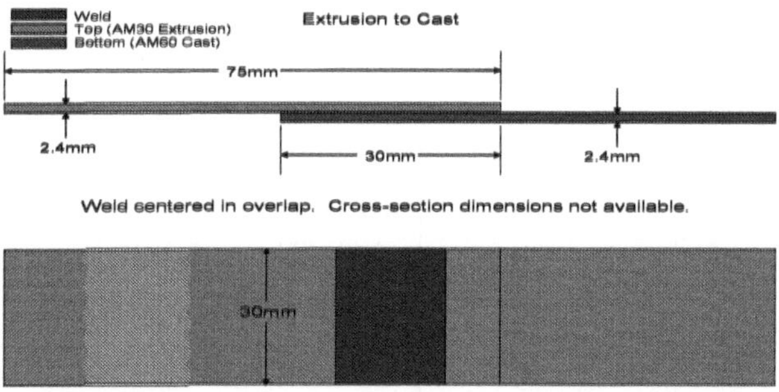

Figure 2. Geometry and dimension of FSW lap-shear specimen for magnesium alloys [12].

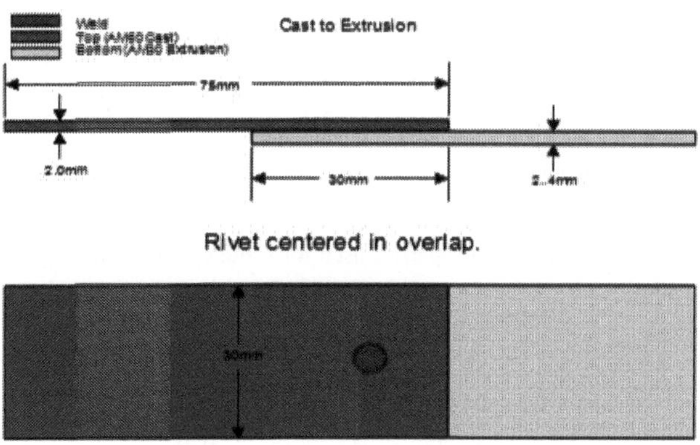

Figure 3. Geometries and dimensions of SPR and SPR with adhesive joint for magnesium alloys [12].

Modeling of FSSW Joints

According to the testing information the specimens were gripped 55 millimeters from the center of the FSSW. Thus, the grip-to-grip distance employed for each specimen was 110 millimeters. Based on the specimen configurations shown in Figure 1, FEA model was created with two dimensional quad type of shell elements (the base material is AZ31 alloy – Sheet). The joining areas of FSSW in both sides of the specimen coupons are circular shape. However, the circular representation of the joining area in FEA model could be inefficient if a vehicle structure contains thousands of FSSW joints. Thus, it was investigated by modeling the weld as circular and square shape as if the shape representation of the weld has any influence on the FEA results and ultimately the fatigue life prediction. It was also studied the effect of the element size with 1 mm, 3 mm, and 5 mm at the peripheral of a FSSW joint.

Table 1. Specimen types and configurations for FSW, SPR, and SPR with adhesive joint

Specimen Type	Joint Type	Base Alloy (Thickness, mm)	Load Ratio
Lap-Shear	FSW	AM30 (2.4) to AM60 (2.4)	R = 0.1
		AZ31 (2.0) to AM60 (3.2)	R = 0.1
		AZ31 (2.0) to AM60 (3.2)	R = 0.3
	SPR	AM30 (2.5) to AM60 (2.0)	R = 0.1
		AM60 (2.0) to AM30 (2.5)	R = 0.1
		AM60 (2.0) to AZ31 (2.0)	R = 0.1
		AM60 (2.0) to AZ31 (2.0)	R = 0.3
		AZ31 (2.0) to AM60 (2.0)	R = 0.1
	SPR with Adhesive	AM30 (2.5) to AM60 (2.0)	R = 0.1
		AM60 (2.0) to AM30 (2.5)	R = 0.1
		AM60 (2.0) to AZ31 (2.0)	R = 0.1
		AM60 (2.0) to AZ31 (2.0)	R = 0.3
		AZ31 (2.0) to AM60 (2.0)	R = 0.1

Table 2. Material properties used for finite element analyses

Material \ Properties	Elastic Modulus (MPa)	Poisson's Ratio	Density (gm/m^3)
AZ31	45,000	0.35	1.88
AM30	45,000	0.35	1.88
AM60	39,000	0.35	1.88

In FEA model an FSSW joint can be represented with various modeling techniques including bar element, ACM (area contact method) connection, and TIE contact condition. For bar element representation of an FSSW joint in FEA model, a bar element connects both sides of the welded area of top and bottom plates represented with 2-D shell elements. The length of the bar element for the FSSW joint is the average thickness of the top and bottom plate. The bar element is located at the center of the welded joint for both plates as shown in Figure 4.

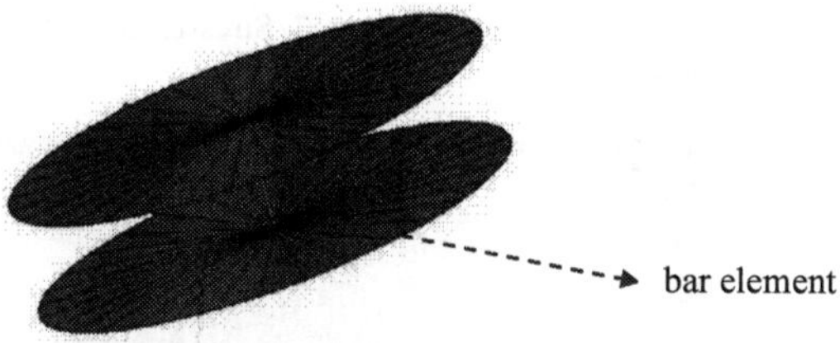

Figure 4. Bar element representation for an FSSW joint.

ACM is a special representation in FEA model for spot welds in vehicle structures. This modeling technique is consisted of number of rigid elements (RBE3) and a solid (HEXA) element whose cross-sectional area is equivalent to the area of the weld nugget. The solid element represent the FSSW and the corner nodes of the HEXA element are connected to top and the bottom sheet using RBE3 element as shown in Figure 5.

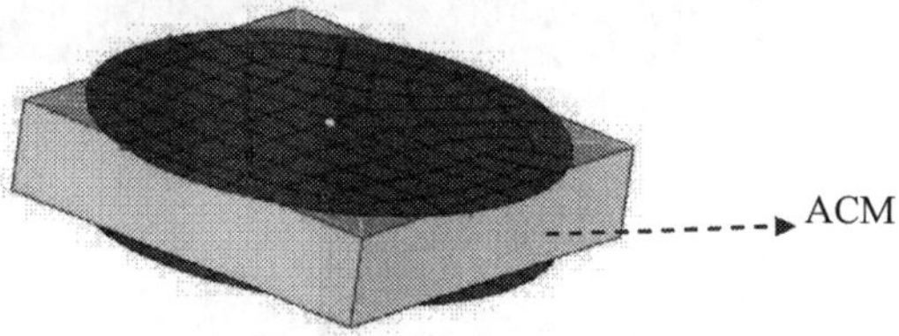

Figure 5. ACM representation for an FSSW joint.

TIE contact condition simulates the behavior of the FSSW joint connecting the two plates. The elements within the weld region of the top and the bottom plate are selected and assigned for contact surfaces. The elements in the contact surface of the bottom plate are assigned as the master surface and the elements in the contact surface of the top plate are assigned as the slave surface. Figure 6 shows an example of the TIE contact condition. No physical element represents the FSSW joint in the model.

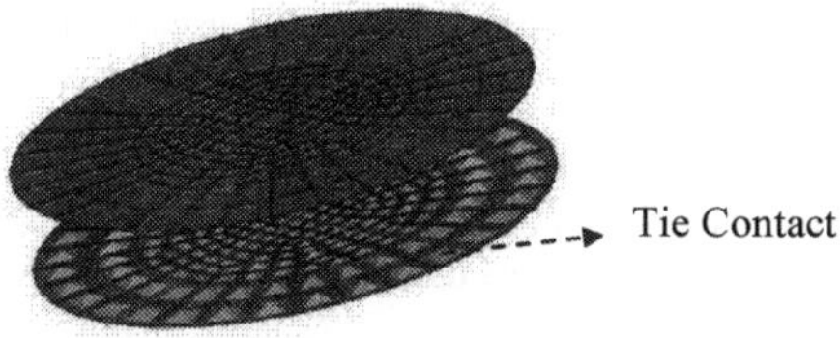

Figure 6. TIE contact representation for an FSSW joint.

Using TIE contact representation for an FSSW joint, FEA models were created for two different representations of the contact surface shapes, which were circular and square. Each contact surface shape was meshed with element sizes of 1 mm, 3 mm and 5 mm. For the square contact surface the area of the square was equal to that of the circular contact surface. The circular and square contact surfaces are shown in Figure 7. Stresses are obtained for the first row of elements which are marked in red.

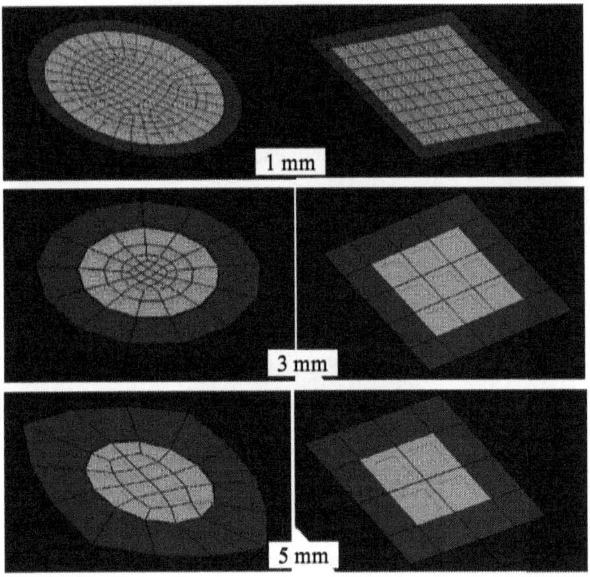

Figure 7. Two different contact surface shapes with three different element sizes.

Modeling of FSW Joints

Specimen geometry and dimensions are shown in Figure 2 and the joint configurations were shown in Table 1 for FSW joints. FSW joints were modeled with TIE contact condition for the sake of the simplicity of modeling. However, the base material was modeled differently depended on the manufacturing process that was used to obtain the base materials. In vehicle industry all casted parts are modeled with 3-D elements since the shape of the parts are generally too complicate to extract mid-surface for 2-D element meshing. Thus, cast magnesium alloy, AM60B, was modeled with 3-D elements. However, sheet and extrusion parts were modeled with 2-D elements. Therefore, a coupon with 2-D elements were connected to the other coupon with 3-D elements for the specimens fabricated with AZ31 to AM60B and AM30 to AM60B.

The models of the specimens were constrained as the physical specimens were gripped in the testing machine. Finite element models were developed with the mesh size of 1, 3 and 5 for all specimen configurations. The coupon represented with 2-D elements was meshed with quad type shell elements. On the other hand, the coupon represented with 3-D elements was meshed with tetra type of 2nd order elements. Spacers at both grip areas were also modeled as 2-D or 3-D elements that depend on the element type of the matching coupon. Figure 8 shows the FEA model created with 2-D elements for AM30 and 3-D elements for AM60B.

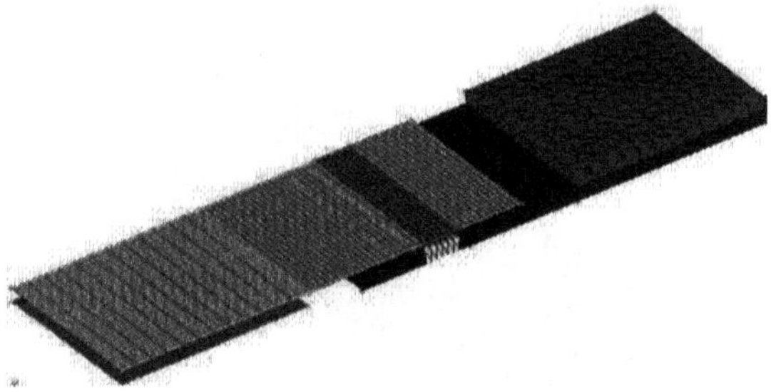

Figure 8. A FEA model meshed with 2-D and 3-D elements for FSW joint specimen.

Material properties of the weld regions are the same with those of the base materials shown in Table 2. Rigid elements were created for the elements within the gripped area. The center node within the gripped regions is assigned as an independent node and all the remaining nodes in the regions were assigned as dependent nodes constrained for all degrees of freedom. In the loading side of the grip the independent node was constrained in all directions except the loading direction which is along the specimen axial direction. The other side of the grip was constrained in all the directions as in the testing machine.

Modeling of SPR Joints

The specimen geometries and dimensions were shown in Figure 3, and the specimen configurations were shown in Table 1 for SPR joints used in this study. The modeling techniques used for SPR joints were the same with those used for FSW joints. However, only two different mesh sizes of 3 mm and 5 mm were investigated. SPR joint in FEA model is represented with TIE contact condition on the square region as shown in Figure 9. The area of the square region is the same with that of the cross section area of the circular SPR. Slight difference in stress analysis results was observed when square representation, instead of circular representation, was used for the cross section of SPR. However, the square representation is simpler than the circular representation in a vehicle structure. Thus, the square representation will be more efficient way of meshing for a structure level.

Modeling of SPR Joints with Adhesive

The same modeling techniques used for FSW and SPR joint specimens were applied to construct the FEA models of the SPR joint with adhesive. However, the adhesive layer between two coupons was represented with ACM connection. Four HEXA8 elements are located between the two coupons and connected to the elements on the coupons from the corner nodes of the HEXA8 elements using RBE3 elements as shown in Figure 10.

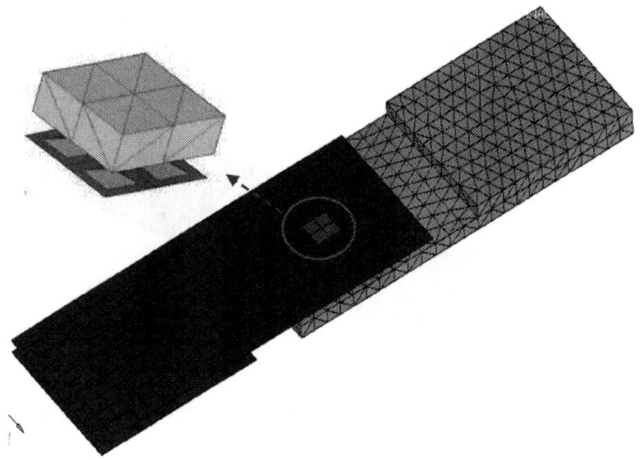

Figure 9. A FEA model meshed with 2-D and 3-D elements for SPR joint specimen.

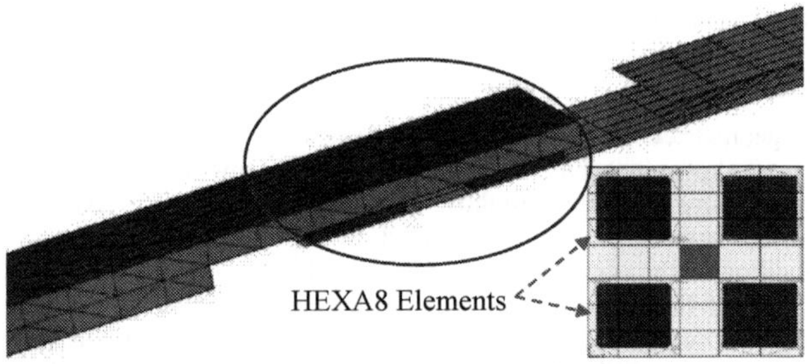

Figure 10. An FEA model for an SPR joint with adhesive bonding.

Stress - Life Curves for Fatigue Life Prediction of the Various Joint Types

Fatigue life prediction of various joints in a vehicle structure can be performed with various methods including stress and strain based approaches. This study focuses on the stress based method to predict fatigue life of the various joints for magnesium alloys. The stresses obtained from linear elastic FEA are correlated to the number cycles to failure of the specimens tested for each joint type.

Stress-Life Curves for FSSW Joints

The most widely accepted method for fatigue prediction of electrical resistance spot welded joints is structural stress methods [13-15]. In this method, forces and moments at the critical locations are obtained to calculate the structural stresses. The same approach can be employed to predict the fatigue life of FSSW joints of magnesium alloys investigated in this study. Most structural stress methods were developed using 2-D shell elements for representing the coupons joined. As described in previous section, casted components are modeled with 3-D elements instead of 2-D elements. This means that there is limitation for expanding the current structural stress methods to FSSW joints composed of casted magnesium alloys. Thus, this study proposed simple stress based fatigue prediction method that can be applied for all kinds of representations in FEA models for the magnesium alloys.

The stress based fatigue life prediction method mainly depends on the approach to obtain the stress at the critical location of the joint. Therefore, in this study various modeling techniques (bar element, ACM connection, and TIE contact condition) were discussed to represent the FSSW joint in FEA model. The maximum stress on the element at the vicinity of ACM connection and TIE contact condition can be correlated with the number of cycles to failure obtained from experiments. For the bar element connection case, structural stress can be calculated using forces and moments at the connecting nodes between the bar and the shell elements. Please note that for the FSSW joint only TIE contact condition was used.

Figure 11 shows the stress distributions of the FSSW specimen subjected to unit load, where the FSSW joint is represented with TIE contact condition, and modeled with three different mesh sizes of 1 mm, 3 mm, and 5 mm. Figure 12 shows the fatigue test results of FSSW joint specimens made of AZ31. The test results are presented with the applied load range versus the number of cycles to failure. Even though fatigue tests were conducted with three load ratios (R= min load/max. load) , R=0, 0.3, and 0.7, no mean load effect is observed. Figure 13 shows the variation of the stress-life curve due to the mesh size differences in FEA models. As shown in the figure, stress ranges with the mesh size 1 mm are higher than those of the mesh sizes of 3 mm and 5 mm. No noticeable differences are observed between the mesh size of 3 mm and 5 mm. Figure 14 shows the stress-life curves that present stress range versus the number of cycles to failure for different cross section representations of the FSSW joint. The circular representation of the FSSW shows slightly higher stress range than that of the square representation at the specimen subjected to the same load range. Both stress ranges were obtained from FEA models with mesh size of 3 mm.

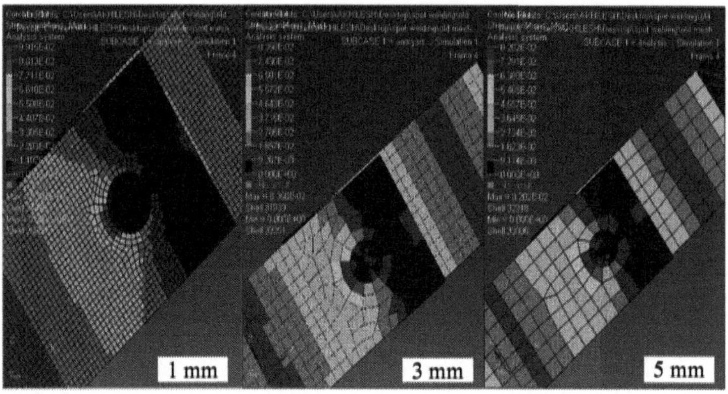

Figure 11. Stress distributions of FSSW specimens subjected to unit load.

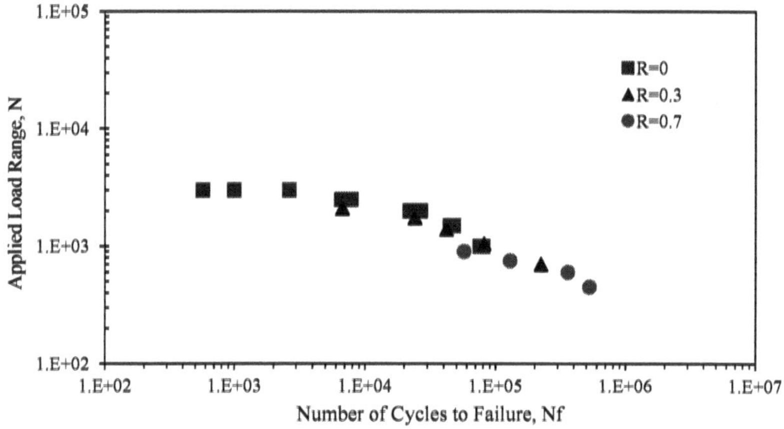

Figure 12. Fatigue test results of FSSW joints for AZ31 magnesium alloys.

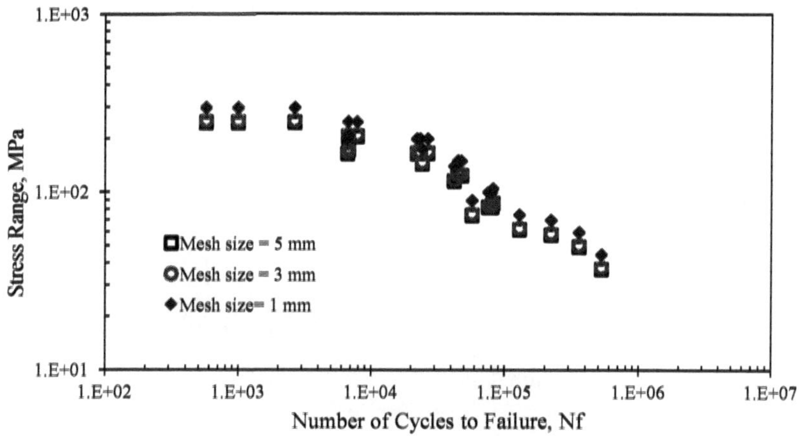

Figure 13. The effect of mesh size on the stress-life curve of FSSW joints.

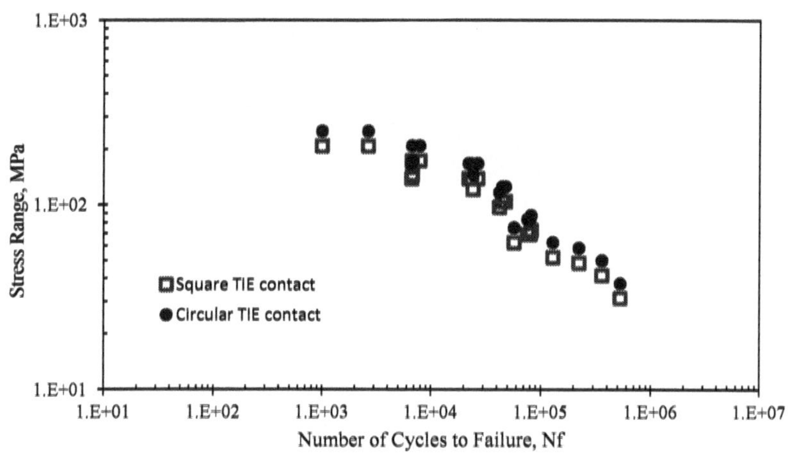

Figure 14. The effect of representation shape of FSSW on the stress-life curve.

Stress-Life Curves for FSW Joints

FSW joint is represented as TIE contact condition between two coupons joined. The maximum stress at the vicinity of the joint representation is correlated to the number of cycles to failure obtained from the laboratory testing. The stress distributions at the FSW specimen subjected to unit load are shown in Figure 15. The mesh sizes used are 3 mm and 5 mm, which is most common element sizes used for modeling in a passenger vehicle structure. Figure 16 shows the stress-life curves for AM60 to AZ31 joining with two different mesh sizes. There is a slight difference as expected in the stress-life curves for the different mesh sizes. Figure 17 shows the stress-life curves for all FSW joint specimens meshed with element size of 3 mm.

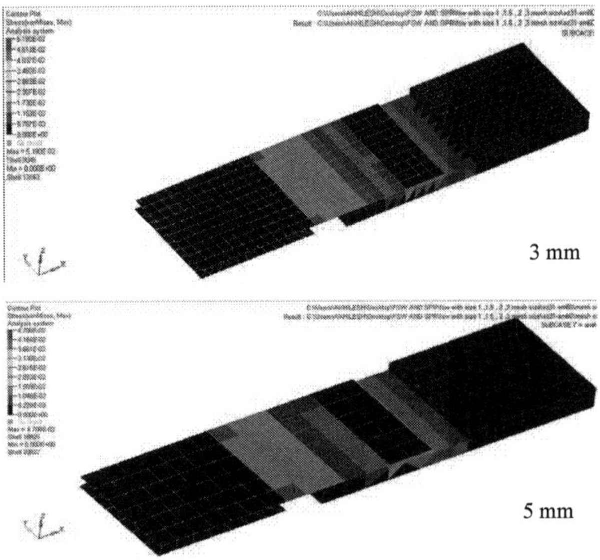

Figure 15. Stress distributions of FSW specimens subjected to unit load condition.

Stress-Life Curves for SPR Joints

SPR joint is represented as TIE contact condition between two coupons joined. The maximum stress at the vicinity of the joint is correlated to the number of cycles to failure obtained from the laboratory testing. The contact areas on the coupons were represented with circular and square shapes in FEA models to characterize the difference in representation of the SPR cross section. The results were similar to those obtained for FSSW as shown in Figure 14. Thus, SPR joint was represented with square elements with two different sizes of 3 mm and 5 mm. Figure 18 shows the stress-life curves for SPR joint specimens (tested with R=0.1) modeled with mesh size of 3 mm. The same trend with Figure 16 was observed for the mesh size of 5 mm of SPR joint specimens. All data points are collapsed within a single narrow band for four specimen configurations.

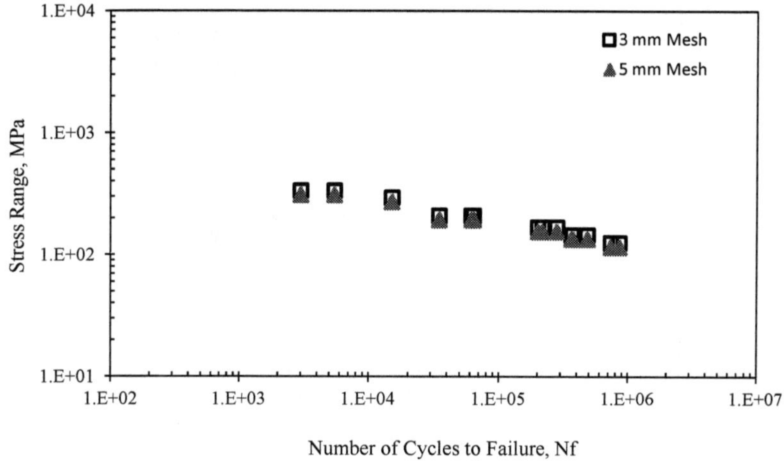

Figure 16. Stress-life curves for AM60-AZ31 specimens with two different mesh sizes.

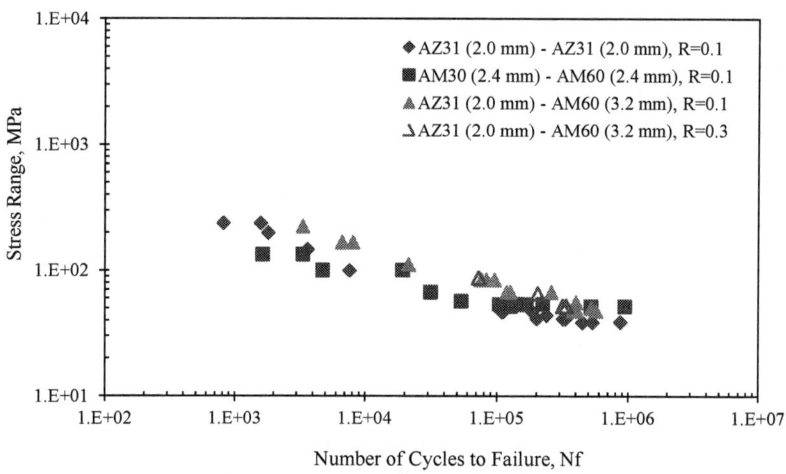

Figure 17. Stress-life curves of FSW specimens.

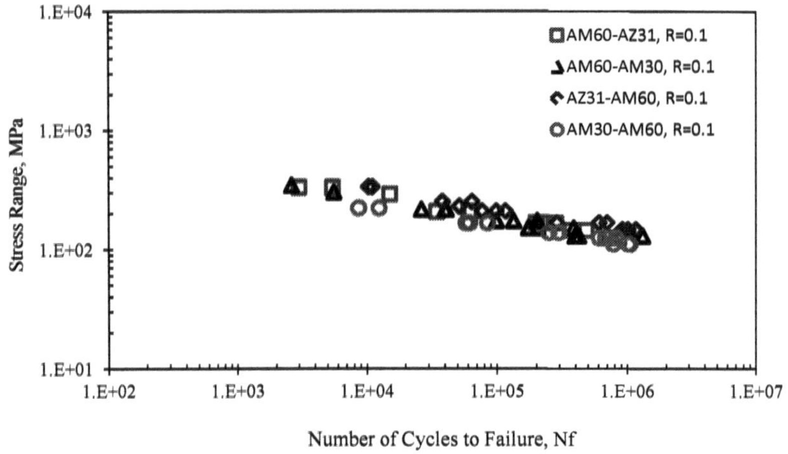

Figure 18. Stress-life curves for SPR joint specimens with mesh size of 3 mm.

<u>Stress-Life Curves for SPR Joint with Adhesive</u>

The SPR joint with adhesive was modeled with the combination of TIE contact condition and ACM connection. The SPR joint is modeled with TIE contact and the adhesive area is modeled with four ACM connections. The maximum stress at the vicinity of the joint is correlated to the number of cycles to failure obtained from the laboratory testing. The same specimen configuration was modeled with mesh size of 3 mm and 5 mm to investigate the effect of mesh size on the stress calculation at the critical area of the specimen. Figure 19 shows the stress-life curves for SPR joint with adhesive. There is a slight difference in slope of the stress-life curves for different mesh sizes. All data points are collapsed within a single narrow band except AM60-AZ31 tested with R=0.3 that showed mean stress effect on the plot.

Discussions, Conclusions and Next Steps

In this study various magnesium alloy joints were modeled with various FEA modeling techniques by considering the application to a vehicle structure. Thus, modeling simplicity with reasonable accuracy was a main factor to employ an FEA modeling technique for the various joints. It was also discussed that the effect of the element sizes at the vicinity of the joints on the FEA results. The element size of 1 mm was resulted in a too big FEA model size, and the element size of 3 mm and 5 mm showed almost the same results in stress-life curves. However, the FEA model with element size of 5 mm showed inferior correlation in stress-life curve of SPR with adhesive joint specimens.

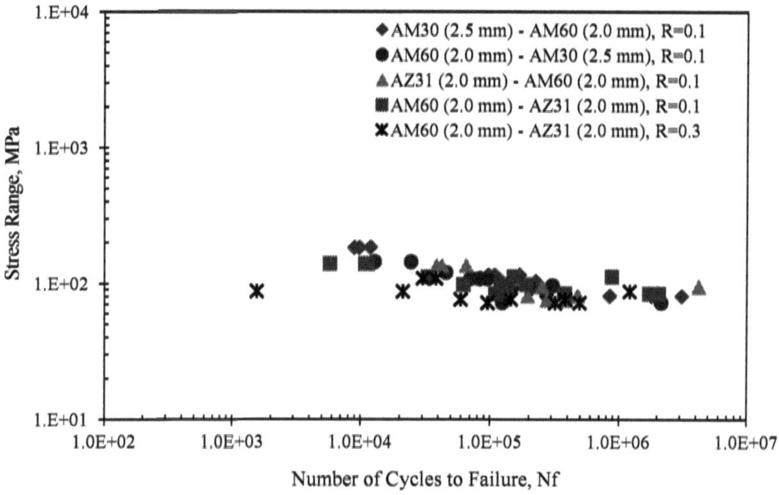

Figure 19. Stress-life plots for the SPR joint with adhesive specimens meshed with 3 mm element.

Stress based fatigue prediction method for various joints of magnesium alloys was discussed. The key for this approach is representation of the joints in FEA models of the various specimen configurations. It is also important to be consistent with the same modeling technique for the joint in the structure and S-N data used for life prediction of the joints. Thus, this study focuses on developing suitable simulation techniques that can be simple to use but can provide reasonable accuracy. The modeling approach used in this study showed the effectiveness for testing specimens, and it may need to be verified for large structures such as a passenger vehicle body structure. This approach inherits the limitations of FEA. For an example the stress values obtained from linear static FEA are always depended on the mesh quality and sizes, which results in alternating the slope of the stress-life curves of the joint configurations. Fortunately, FEA model for a passenger vehicle generally contains the element size between 3 mm to 5 mm. Nevertheless, it is necessary to verify this stress-life approach with structural components contain the various joints discussed in this study. The conclusions of this study are:

- The representation of the joints with TIE contact eliminates the time consuming modeling process for 3-D meshing for the joints. It is also very effective to represent the joint specimens modeled with 2-D elements for one coupon and 3-D elements for the other coupon.

- For the future study, the stress-life curves obtained from this study will be employed for the fatigue life predictions of a sub-structure of a passenger vehicle that contains FSW and SPR joints.

References

1. G.S. Cole and A.M. Sherman, "Lightweight Materials for Automotive Applications," *Materials Characterization*, 35 (1995), 3–9.
2. L. K. France and R. Freeman, "Welding and Joining of Magnesium," SAE Technical Paper, 2001-01-3443, 2001.
3. K. Okamoto, F. Hunt and S. Hirano, "Friction Stir Welding of Magnesium for Automotive Applications," SAE Technical Paper 2005-01-0730, 2005.
4. L. Agarwal, P.K. Mallick and H.T. Kang, "Spot Friction Welding of Mg-Mg, Al-Al and Mg-Al Alloys, SAE Technical Paper, 2008-01-0144, 2008.
5. P.K. Mallick and L. Agarwal, "Fatigue of Spot Friction Welded Joints of Mg-Mg, Al-Al and Al-Mg Alloys," SAE Technical Paper, 2009-01-0024, 2009.
6. S. Lathabai, V. Tyagi, D. Ritchie, T. Kearney, B. Finnin, S. Christian, A. Sansome, and G. White, "Friction Stir Blind Riveting: A Novel Joining Process for Automotive Light Alloys," SAE Technical Paper, 2011-01-0477, 2011.
7. P.K. Mallick and S. Bhambure, "Analysis of Single Lap Adhesive Joints between Magnesium and Other Structural Automotive Materials," SAE Technical Paper, 2011-01-0076, 2011.
8. W.M. Thomas, E.D. Nicholas, J.C. Needham, M.G. Murch, P. Temple-Smith, and C.J. Dawes, "Friction-stir butt welding," GB Patent No. 9125978.8, International patent application No. PCT/GB92/02203, (1991).
9. G. S. Booth, C.A. Olivier, S.A. Westgate, F. Liebrecht, and S. Braunling, "Self-Piercing Riveted Joints and Resistance Spot Welded Joints in Steel and Aluminium, SAE Technical Paper, 2000-01-2681, 2000.
10. M. Fu and P.K. Mallick, "Effect of Process Variables on the Static and Fatigue Properties of Self-Piercing Riveted Joints in Aluminum Alloy 5754," SAE Technical Paper, 2001-01-0825, 2001.
11. J.B. Jordan, M.F. Horstemeyer, S.R. Daniewicz, H. Badarinarayan, and J. Grantham, "Fatigue Characterization and Modeling of Friction Stir Spot Welds in Magnesium AZ31 Alloy," *Journal of Engineering Materials and Technology*, 132 (2010), 041008-1-10.
12. X. Su, A.K. Khosrovaneh, L. Zhang, Y.-L. Lee, B. Jordon, H.T. Kang, C. Jiang, R. Tagore, and M.F. Horstemeyer, "MAGNESIUM FRONT END RESEARCH AND DEVELOPMENT (MFERD): A Canada-China-USA Collaborative Research & Development Project," (USCAR Annual Report for Task 2.3 Fatigue & durability, 2012).
13. P. Dong, "A Structural Stress Definition and Numerical Implementation for Fatigue Analysis of Welded Joints," *International Journal of Fatigue*, 23 (2001), 865-876.
14. M. Fermér, M. Andréasson and B. Frodin, "Fatigue Life Prediction of MAG-Welded Thin-Sheet Structures," SAE Technical Paper 982311, 1998.
15. H.T. Kang, A.K. Khosrovaneh, M. Amaya, J. Bonnen, H.-C. Shih, S. Mane, and T. Link, "Application of Fatigue Life Prediction Methods for GMAW Joints in Vehicle Structures and Frames," SAE Technical Paper, 2011-01-0192, 2011.

Fatigue of Materials II Advances and Emergences in Understanding
Edited by: T.S. Srivatsan, M. Ashraf Imam, and R. Srinivasan
TMS (The Minerals, Metals & Materials Society), 2013

FATIGUE BEHAVIOR OF AM60B SUBJECTED TO VARIABLE AMPLITUDE LOADING

Kang, H.[1], Kari, K.[1], Khosrovaneh, A.K.[2], Nayaki, R.[1], Su, X.[3], Zhang, L.[4], Lee, Y.-L[4].
[1]The University of Michigan-Dearborn, 4901 Evergreen Road, Dearborn, MI 48128
[2]General Motors Holdings, Warren, MI
[3]Ford Motor Corporation, Dearborn, MI
[4]Chrysler Corporation LLC, Auburn Hills, MI

Keywords: Fatigue, Variable Amplitude Loading, AM60B

Abstract

Magnesium alloys are considered as an alternative material to reduce vehicle weight due to their weight which are 33% lighter than aluminum alloys. There has been a significant expansion in the applications of magnesium alloys in automotives components in an effort to improve fuel efficiency through vehicle mass reduction. In this project, a simple front shock tower of passenger vehicle is constructed with various magnesium alloys. To predict the fatigue behavior of the structure, fatigue properties of the magnesium alloy (AM60B) were determined from strain controlled fatigue tests. Notched specimens were also tested with three different variable amplitude loading profiles obtained from the shock tower of the similar size of vehicle. The test results were compared with various fatigue prediction results. The effect of mean stress and fatigue prediction method were discussed.

Introduction

For the past decade, the Automotive industry has been engaged on improving the fuel efficiency through vehicle mass reduction and other factors. Currently, magnesium alloys which are 33% lighter than aluminum alloys are considered as an alternative material to reduce vehicle weight [1, 2]. There has been a noticeable expansion in the applications of magnesium alloys in automotives components such as powertrain, chassis and body. Two magnesium alloys; AM50 and AM60 are widely used in door inners and instrument panel beams and the applications of these materials are growing in automotive components (especially for the components and structures required better ductility) [3]. Another magnesium alloy that is widely used is AZ91 which is mainly applied to the powertrain applications, brake pedals, and mounting brackets [3].

This study investigated the fatigue characteristics of AM60B subjected to constant strain amplitudes and variable amplitudes. Fatigue life of AM60B under the variable amplitude loadings was predicted using Smith-Watson-Topper (SWT) equation with fatigue properties obtained from the strain controlled fatigue tests. Strain amplitudes for fatigue life predictions were obtained from finite element analysis (FEA) with two different element mesh sizes, 3 mm and 5mm. Finally the fatigue test results of AM60B subjected to the variable amplitude loadings were compared with the fatigue life prediction results obtained from Smith-Watson-Topper equation.

Experimental Procedure

Fatigue Properties of AM60B

Strain controlled fatigue testing was conducted with various strain amplitude to obtain the fatigue properties ($\sigma'_f, b, \varepsilon'_f, c$) of AM60B. The procedures to determine these properties were well described in [4]. The fatigue strength coefficient (σ'_f) and the fatigue strength exponent (b) were obtained from Basquin's equation as presented:

$$\frac{\Delta\sigma}{2} = \sigma'_f (2N_f)^b \tag{1}$$

where $\frac{\Delta\sigma}{2}$ = true stress amplitude, $2N_f$ = reversals to failure (1 rev = ½ cycle). The true stress amplitude versus the reversals to failure was plotted in log-log scale. The intercept of the least squares fit line is the fatigue strength coefficient (σ'_f) and the slope of the line is the fatigue strength exponent (b). The stress amplitude can be replaced with elastic strain amplitude:

$$\frac{\Delta\varepsilon_e}{2} = \frac{\Delta\sigma}{2E} \tag{2}$$

Thus, the equation (1) becomes:

$$\frac{\Delta\varepsilon_e}{2} = \frac{\sigma'_f}{E} (2N_f)^b \tag{3}$$

Where $\frac{\Delta\varepsilon_e}{2}$ = elastic strain amplitude, and $\frac{\Delta\sigma}{2}$ = stress amplitude. The intercept of the least squares fit line becomes $\frac{\sigma'_f}{E}$ and the slope of the line is the fatigue strength exponent (b). The elastic strain amplitude versus the number of reversals to failure for AM60B is shown in Figure 1. From the plot fatigue strength exponent (b = slope of the curve = -0.205) is determined, and the intercept of the curve with the vertical axis, $\frac{\sigma'_f}{E}$ = 0.0213, is also determined. The fatigue strength coefficient can be calculated with the monotonic modulus of elasticity of AM60B.

The fatigue ductility coefficient (ε'_f) and the fatigue ductility exponent were obtained from Coffin and Manson equation:

$$\frac{\Delta\varepsilon_p}{2} = \varepsilon'_f (2N_f)^c \quad (4)$$

where $\frac{\Delta\varepsilon_p}{2}$ = plastic strain amplitude, $2N_f$ = reversals to failure (1 rev = ½ cycle), ε'_f = fatigue ductility coefficient, and c = fatigue ductility exponent. The plastic strain amplitude versus the reversals to failure was plotted in log-log scale. The intercept of the least squares fit line is the fatigue ductility coefficient (ε'_f) and the slope of the line is the fatigue ductility exponent (c).

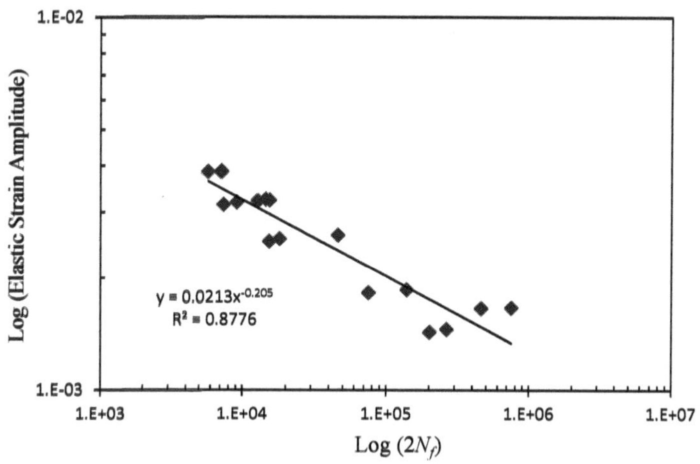

Figure 1. A log-log plot to determine fatigue strength coefficient and exponent.

The plastic strain amplitude ($\frac{\Delta\varepsilon_p}{2}$) was calculated from the measured total strain amplitude, stress amplitude, and the monotonic modulus of elasticity (E) of the AM60B since the total strain amplitude consists of elastic strain amplitude and plastic strain amplitude:

$$\frac{\Delta\varepsilon_p}{2} = \frac{\Delta\varepsilon_t}{2} - \frac{\Delta\varepsilon_e}{2} = \frac{\Delta\varepsilon_t}{2} - \frac{\Delta\sigma}{2E} \quad (5)$$

Where $\frac{\Delta\varepsilon_t}{2}$ = total strain amplitude, $\frac{\Delta\varepsilon_e}{2}$ = elastic strain amplitude, and $\frac{\Delta\sigma}{2}$ = stress amplitude. The plastic strain amplitude versus the number of reversals to failure for AM60B is shown in

Figure 2. From the plot the fatigue ductility exponent (c = slope of the curve = -0.668) is determined, and the fatigue ductility coefficient ($\varepsilon'_f = 0.3813$), intercept of the curve with the vertical axis, is also determined.

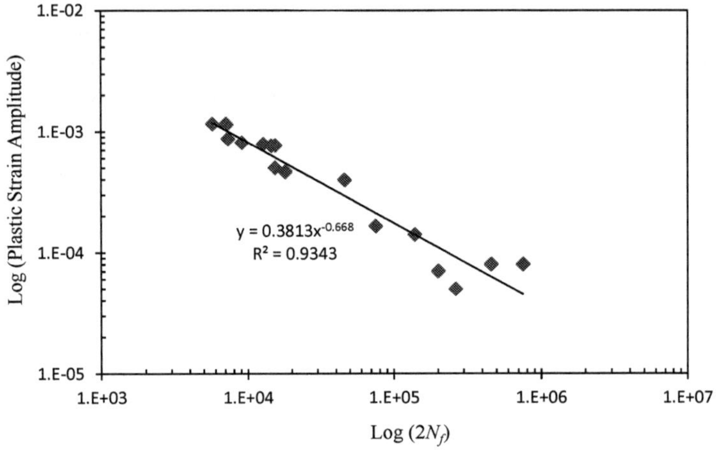

Figure 2. A log-log plot to determine fatigue ductility coefficient and exponent.

From equation 3, 4 and 5, the total strain amplitude can be represented as:

$$\frac{\Delta\varepsilon}{2} = \frac{\sigma'_f}{E}(2N_f)^b + \varepsilon'_f(2N_f)^c \tag{6}$$

This equation is presented with test results of the cast magnesium AM60B in Figure 3.

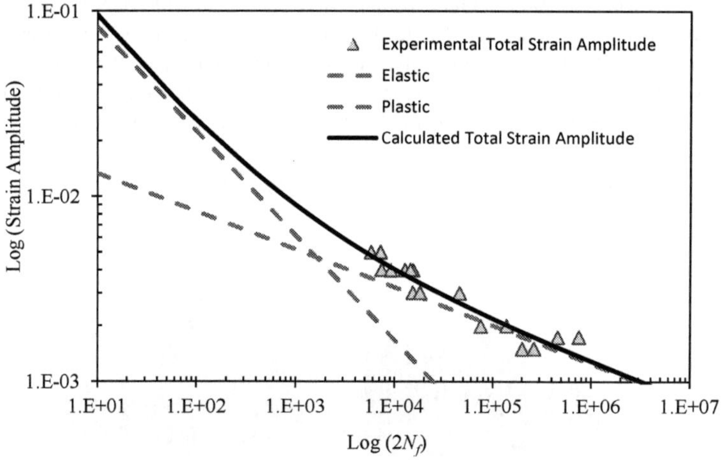

Figure 3. Strain amplitude versus the number of reversals to failure for AM60B.

Fatigue Characteristics of AM60B Subjected To Variable Amplitude Loading

The specimen for fatigue tests of AM60B with variable amplitude loadings is designed considering the stress concentration factor and the critical stress of the material in order to prevent buckling.

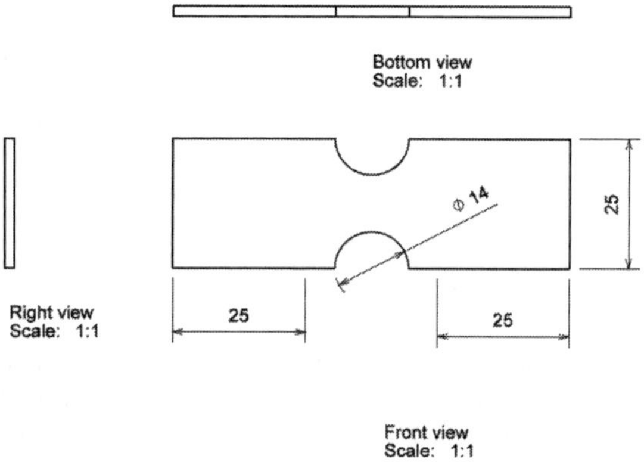

Figure 4. Specimen dimensions in *mm* unit.

The fatigue tests with variable amplitude loadings were conducted with MTS 810 loading frame equipped with TestStarIIs controller. The load profile was obtained at a shock tower of a mid-size passenger vehicle. However, the actual load profile obtained at the shock tower was normalized by the maximum loading event in the load profile as shown in Figure 5. Thus, load amplitudes should be multiplied to the normalized load profile to conduct the fatigue tests of the specimens with variable amplitude loadings. Three different load amplitudes, 3700 N, 4500 N, and 5000 N, were applied to the fatigue tests in this study. Three specimens were tested at the same load profile.

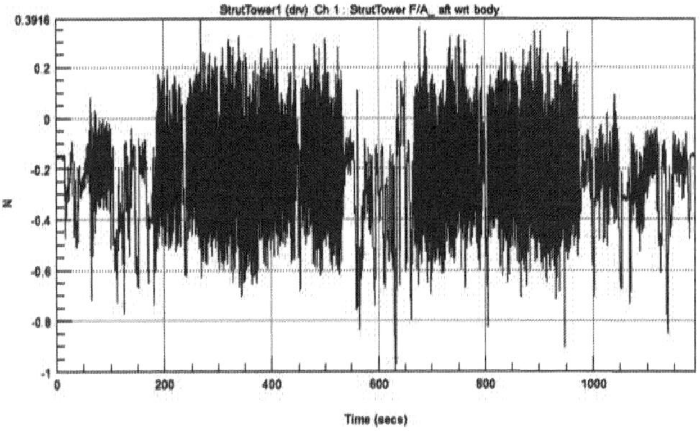

Figure 5. A normalized load profile obtained at a shock tower of a mid-size passenger car.

Fatigue Prediction Of AM60B Subjected To Variable Amplitude Loadings

Fatigue life prediction of AM60B subjected to variable amplitude loadings was conducted with finite element analysis results and fatigue life calculation software called nCode DesignLife. The finite element analysis was conducted with HyperMesh as a pre-processor and Radioss as the solver.

Finite Element Analysis (FEA) Models

Two different FEA models were developed to evaluate the sensitivity of the mesh sizes. The FEA models were constructed with second order tetrahedral solid elements. Those were meshed with 3 mm and 5 mm using a pre-processor called HyperMesh as shown in Figure 6. One side was fully constrained in 6-degrees of freedom and the other side of the specimen was also constrained in all directions except the direction of loading, where 1 N of load was applied.

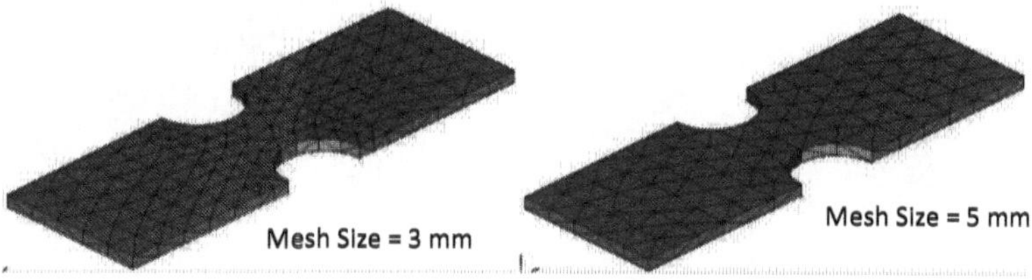

Figure 6. Two FEA mesh sizes for AM60B specimens.

Linear static analysis of the AM60B specimens with 3mm and 5mm mesh sizes were performed in Radioss. The material properties used are the modulus of elasticity, $E = 39000$ MPa, Poisson's Ratio, $\nu = 0.35$, and mass density, $\rho = 1.8\times10^{-9}$ tonne/mm^3. The von Mises stress distributions for each model are shown in Figure 7.

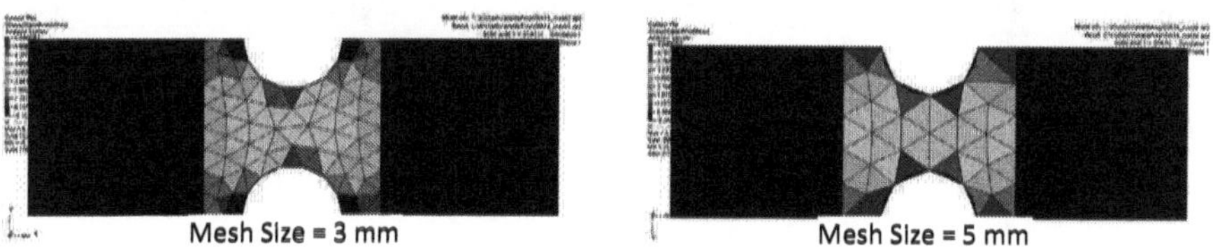

Figure 7. von Mises stress distributions of AM60B specimens subjected to 1 N of loading.

Fatigue Analysis of AM60B

Fatigue analysis of the specimen was conducted with a commercial fatigue software called nCode DesignLife. Fatigue analysis requires the information about the material's fatigue properties, stresses and strains (loading profiles), and fatigue life calculation methods. The fatigue properties of the AM60B were obtained from series of strain controlled fatigue tests, and discussed in the previous section of this paper. The actual stresses/strains on the specimen are obtained from simple multiplication of the FEA outputs of AM60B subjected to the unit load to the variable amplitude load profiles (this will be done internally by the computer code.). The local strain-life approach was used in this study. Neuber's equation was used for elastic-plastic correction. Then, Smith-Watson-Topper method was used for mean stress correction. The fatigue life analysis flow used in nCode DesignLife are shown in Figure 8.

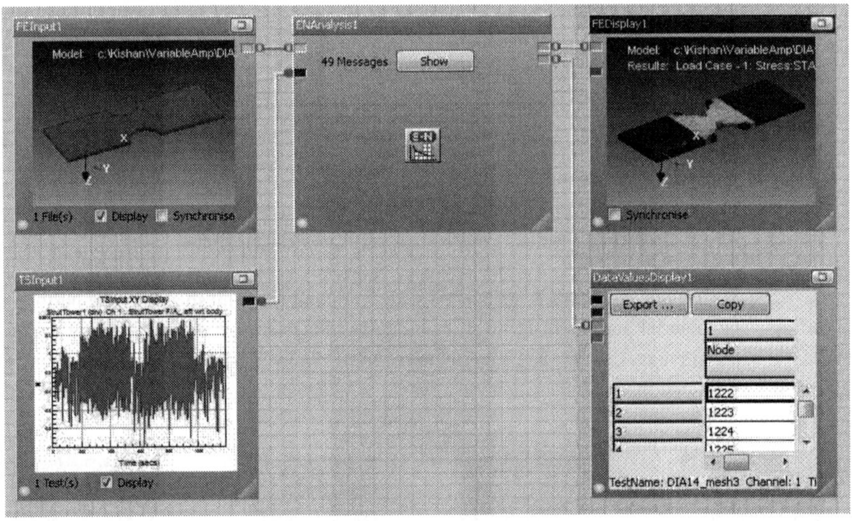

Figure 8. Fatigue life analysis flow used in nCode DesignLife.

Fatigue life analysis results are compared with the experimental results as shown in Figure 9 and 10. The fatigue life calculated using nCode DesignLife with stresses/strains information from mesh size of 3mm in FEA model is presented in Figure 9, and from the mesh size of 5 mm is presented in Figure 10. The strain combination methods used in this analysis were the signed von Mises and the absolute maximum principal strain as shown in both figures. The solid line in both Figures represent the perfect correlation between experiment and prediction results. Two dotted lines at both Figures are factor of 2 lines. As shown in Figures 9 and 10, most of the data points are located within the factor of 2 lines. In the mesh size of 3 mm, the predicted fatigue life well correlates to the test results at the short life region but the predicted one estimates conservatively in the long life region of the data. On the other hand, the predicted fatigue life is shorter than that of test results in short life region but the predicted life well correlates to the test results in the long life region for the mesh size of 5 mm. No noticeable difference in the strain combination methods used is observed as shown in the Figures 9 and 10.

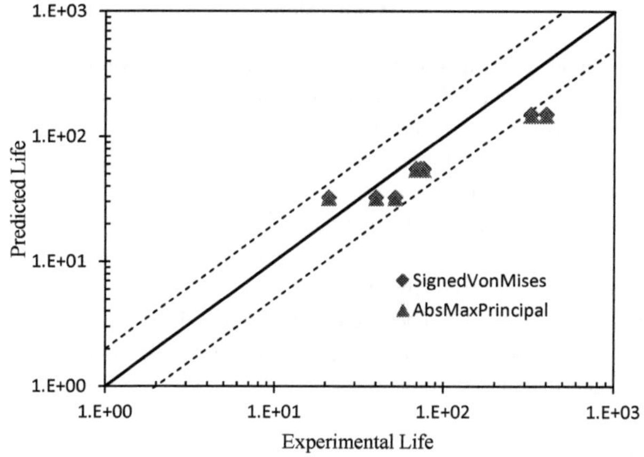

Figure 9. Comparison of fatigue life from prediction with 3 mm mesh size in FEA with experiment results of AM60B.

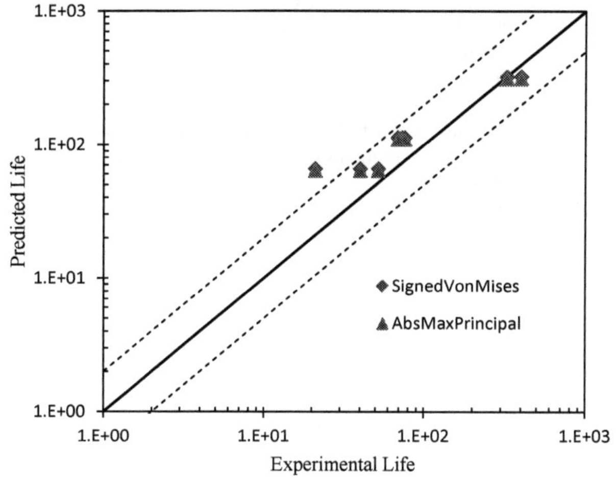

Figure 10. Comparison of fatigue life from prediction with 5 mm mesh size in FEA with experiment results of AM60B.

Conclusions

- This study conducted fatigue tests of a magnesium alloy AM60B subjected to variable amplitude loadings. The following are the conclusion based on this study: Strain-life properties for AM60B was obtained.
- Strain combination methods (sign von Mises and abs max principal) do not make noticeable differences in fatigue analysis results.
- The size of mesh in FEA does affect to fatigue analysis results. This study only employed two mesh sizes, 3 mm and 5 mm, and both models produced acceptable results even though some differences were observed.
- Good correlation was obtained between the test and analysis results.
- The local strain approach which is mainly applicable to aluminum and steel can be also used for magnesium alloys.

References

1. G.S. Cole and A.M. Sherman, "Lightweight Materials for Automotive Applications," *Materials Characterization*, 35 (1995), 3-9.
2. S. R. Agnew, S. Viswanathan, E. A. Payzant, and K. C. Liu, "Tensile and Compressive Creep Behavior of Magnesium Die Casting Alloys Containing Aluminum", THERMEC'2000 - *Proceedings International Conference on Processing and Manufacturing of Advanced Materials*, Las Vegas, USA, December 2000:CDROM, Section A1, Vol 117/3.
3. C.H. Caceres, T. Sumitomo, and M. Veidt, "Pseudoelastic behavior of cast magnesium AZ91 alloy under cyclic loading-unloading," *Acta Materialia*, 51 (2003), 6211-6218.
4. M.L. Roessle, A. Fatemi, and A.K. Khosrovaneh," Variation in Cyclic Deformation and Strain-Controlled Fatigue Properties Using Different Curve Fitting and Measurement Techniques," SAE Technical Paper 1999-01-0364, 1999.

OTHER MATERIALS

Fatigue of Materials II Advances and Emergences in Understanding
Edited by: T.S. Srivatsan, M. Ashraf Imam, and R. Srinivasan
TMS (The Minerals, Metals & Materials Society), 2013

Rubber fatigue – the intrinsic Intricacies

Jean-Louis Poisson, Florian Lacroix, Stéphane Méo, Gaelle Berton, Narayanaswami Ranganathan
Université François Rabelais de Tours
Laboratoire de Mécanique et Rhéologie,CERMEL center,
7 avenue Marcel Dassault, 37200 Tours, France

Keywords: Fatigue, rubber, fractography

Abstract

This paper brings out the salient fractographic features that can be found on the fracture surface of some selected rubbers. It is shown that the presence of a specific feature – called the tongues can be correlated to the effect of load ratio and hence the fatigue lives in a synthetic rubber.

Introduction

Due to their interesting mechanical behavior and their diversity (monomer, formulation), rubber – like materials are more and more used in industry. Understanding phenomenon induced by complex fatigue loadings is an important issue for the conception of industrial components. Therefore, many studies were realized to estimate fatigue lives of rubber parts, in the fatigue crack nucleation frame. Several criteria were developed in that objective:

- First principal stress [1, 2] and strain [3, 4].
- Strain energy density [4, 5]
- Cracking energy density [6, 7]
- Dissipated energy density [8 – 11]
- Eshelby [12, 13]

Nevertheless, these mechanical parameters don't give information about the consequences of fatigue loading on the material itself. Different works were destined to observe some aspects of rubber fracture morphologies. They can be divided into two categories: static and cyclic failures. In this paper, a literature review, detailing both static and cyclic ruptures is first developed. Experimental conditions are then explained, with the material details, and the loading conditions. Finally, fractographic observations are presented and discussed.

Literature review

Bascom [14], one of the pioneers who have explored the failure morphology of rubber – like materials, conceived a specimen holder, which can be installed in a scanning electron microscope (SEM) and is able to maintain a rubber specimen in a bend configuration, so that he can observe the evolution of a crack tip (see Figure 1).

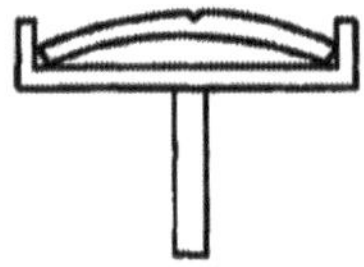

Figure 1. Scheme of specimen holder that hold the specimen in a bend configuration [14]

Precisions about the specimen are given in [15]. He observed the crack tip, created with a razor blade, for different rubber – like materials (nitrile (NBR), polybutadiene (BR), natural rubber (NR)). He noticed fibrous networks and cavities created at the crack tip.

Setua et De [16] investigated on the rupture of SBR occurred with both tensile and tear tests (see Figure 2).

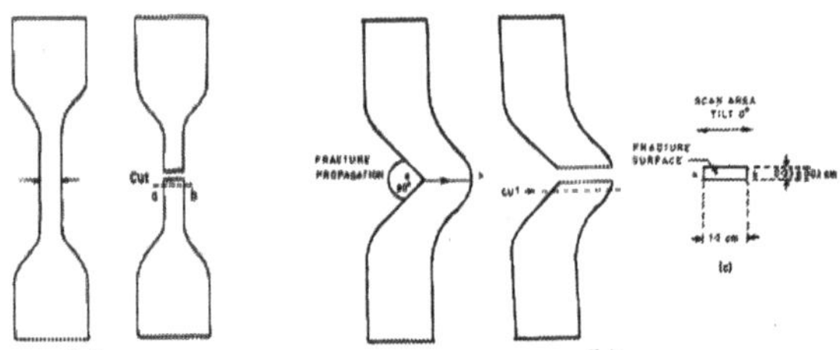

(a) (b)

Figure 2. (a) tensile specimen, (b) 90° tear specimen [16 – 18]

They observed the influence of vulcanization and reinforcement on the morphology of failure surfaces. They remarked differences between rupture of peroxide and sulfure cured SBR. They also pointed out the impact of carbon black on the failure surfaces of the materials. Pandey et al. [19] have studied the static behavior of a BIIR – CR copolymer in tensile and tear tests. They correlated their mechanical behavior with their relative fracture topography. They noticed defects created by the introduction of the fillers (see Figure 3).

Figure 3. Material defect induced by filler detachment [19]

Luong et al. [20] compared the crack growth behavior of a SBR and a NR, with the pure shear specimen (see Figure 4).

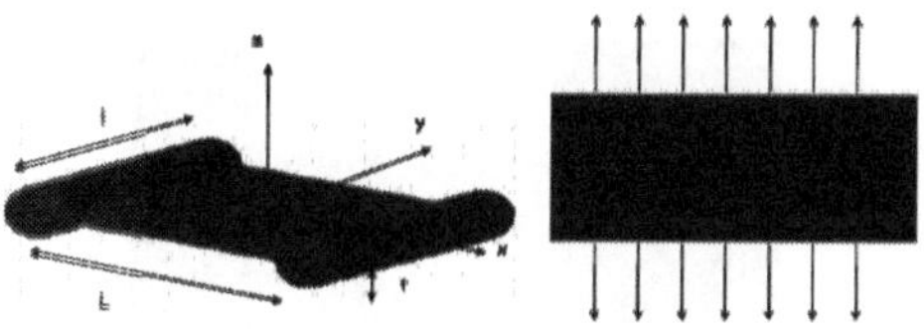

(L=157 mm, l=34 mm and thickness, t=2 mm)

Figure 4. Pure shear specimen [20]

They noticed the creation of small cavitations at the early propagation phase for both materials (see Figure 5).

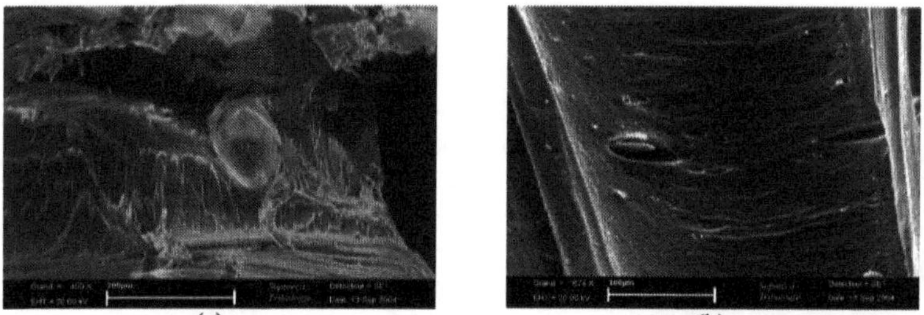

(a) (b)

Figure 5. SEM photomicrographs of early propagation of (a) NR (b) SBR [20]

Bhowmick et al. [17] worked on the influence of the loading on the fracture morphology. They pointed out specificities related to the loading conditions for a NBR, and then for a NR [18]. Consequently, static rupture morphology investigations are not adapted to explain cyclic fatigue phenomena at the material scale. Therefore, Goldberg et al. [21] analyzed the cyclic rupture of SBR tensile specimens. They noticed the impact of different test parameters on the fracture morphologies (temperature, maximum stress). They also divided their rupture surfaces in two regions: initiation and fast fracture zones (see Figure 6)

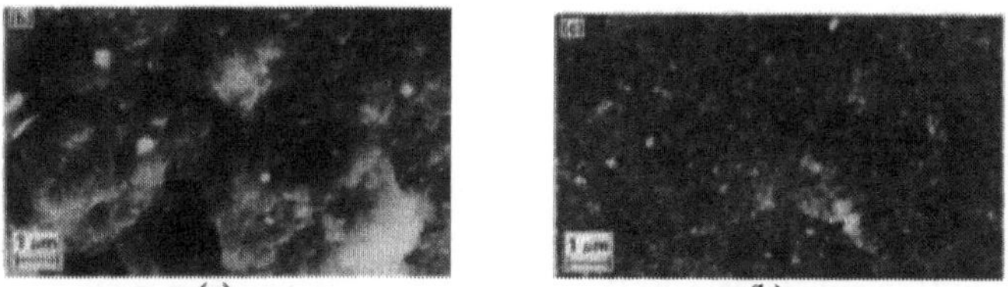

(a) (b)

Figure 6. Microphotographs of fracture surface of a SBR material subjected to cyclic tension loading. (a) initiation zone. (b) fast fracture zone [21]

Hainsworth [22] carried out fatigue tests on tensile specimens applied to a silicone rubber and a NR. She remarked that the crack was initiating to a defect created by the specimen cutting process (see Figure 7).

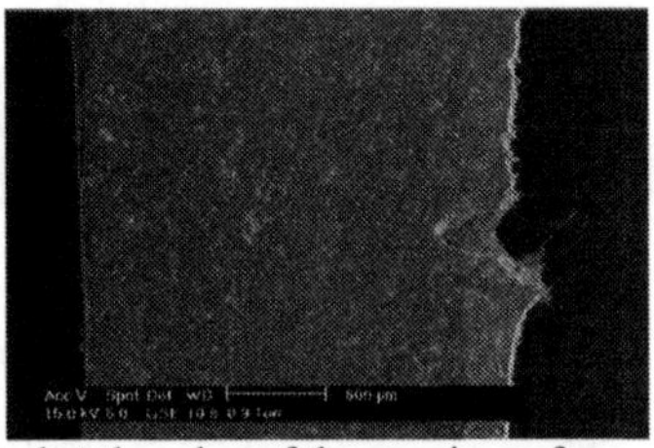

Figure 7. Fatigue crack initiated at the edge of the specimen from a defect created by the cutting process [22]

She also proposed a scheme of the material fatigue damaging, from a macro scale observation of failure surfaces. They are there divided into three regions, with the initiation zone, the ductile zone and the final failure zone (see Figure 8).

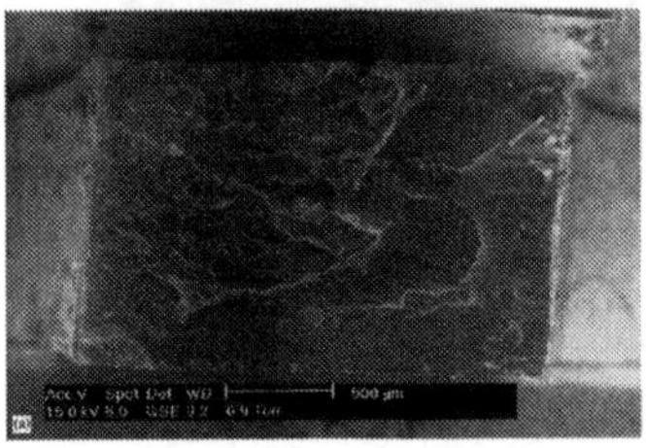

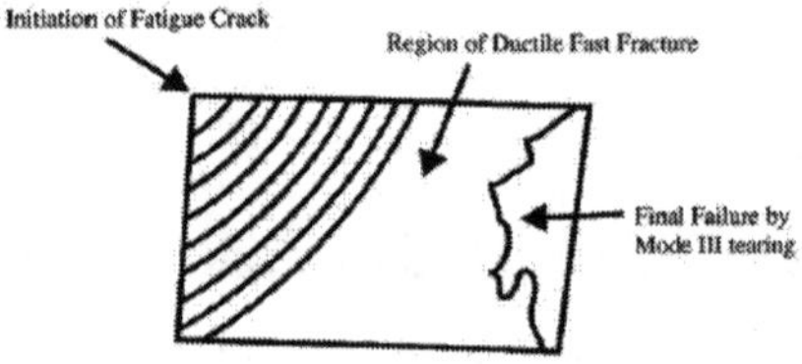

(a) (b)

Figure 8. (a) microphotograph of a macro-scale failure surface (b) Explaining sketch of the corresponding failure surface [22]

This last result indicates that observing the fracture morphologies brings out information on the phenomena induced by fatigue testing. To this end, many works were realized to understand the changes caused on the material by fatigue loadings.

Legorju – Jago [23] realized fractographic investigations on natural rubber crack growth specimens, for different mechanical parameters (maximum strain, frequency, temperature). She noticed the presence of tongues in a zone close to the crack tip, of which the morphology vary with the frequency.

Lacroix [11] observed polychloroprene (CR) dumbbell specimen's failure surfaces, subjected to tension loading. He concluded that the main crack was propagating from micro flaws presents in the virgin material (see Figure 9).

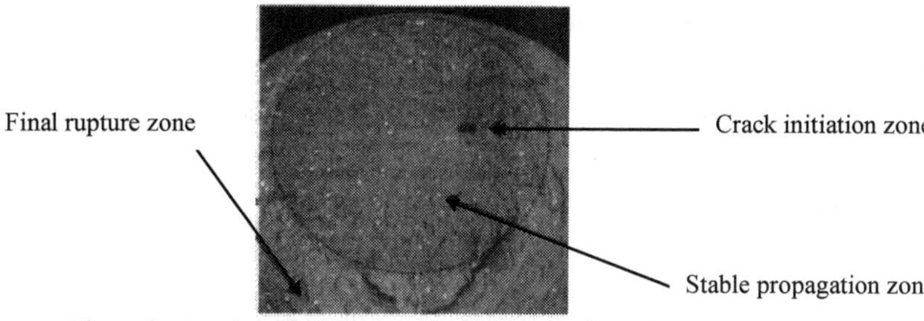

Figure 9. – crack initiation zone of a CR dumbbell specimen [11]

He optimized this compound's fatigue lives by increasing its homogeneity. He also found out tongues, similar to those observed by Legorju – Jago [23], on the crack initiation region (see Figure 10).

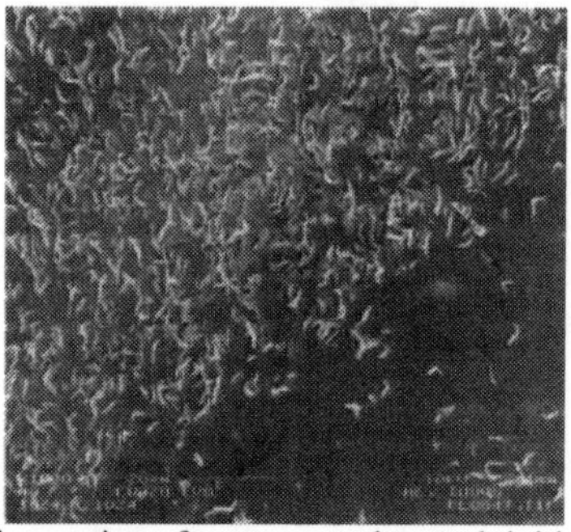

Figure 10. – Observation of tongues on the crack initiation zone [11]

Gauchet [24] analyzed the impact of the type of carbon black on the fatigue damaging of a hydrogenated nitrile (HNBR), and particularly on the failure surface morphology. She defined a petals shaped rough zone (see Figure 11) corresponding to the crack propagation region.

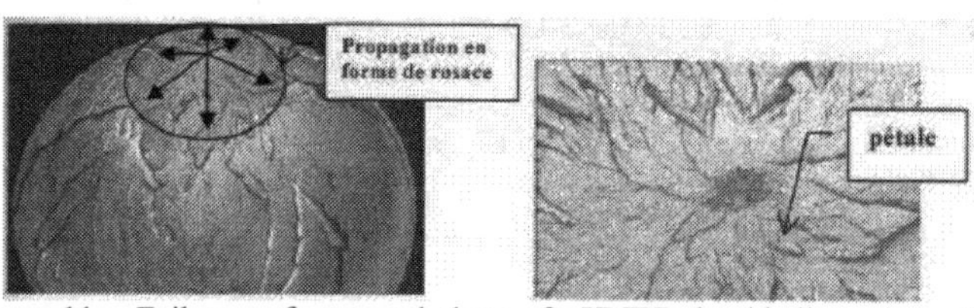

Figure 11. – Failure surface morphology of a HNBR dumbbell specimen [25]

She pointed out the increase of this area with the decrease of carbon black's specific surface. She also highlighted the impact of temperature and stress levels on this crack propagation zone (see Table I).

Table I. Table that summarize the effect of temperature and stress maximum on the stable propagation zone [24]

Temperature	20°C	120°C	Stress	1.3 MPa	1.4 MPa
Stable propagation zone area	++	+	Stable propagation zone area	++	+
Petals size	+	++	Petals size	+	++
Petals number	++	+	Petals number	++	+
Tongues density	+	++	Tongues density	+	++

Le Cam et al. [26] proposed to identify the influence of the strain ratio on the fatigue rupture of NR specimens subjected to tension loading. They defined three regions, depending on the strain ratio (see Figure 12).

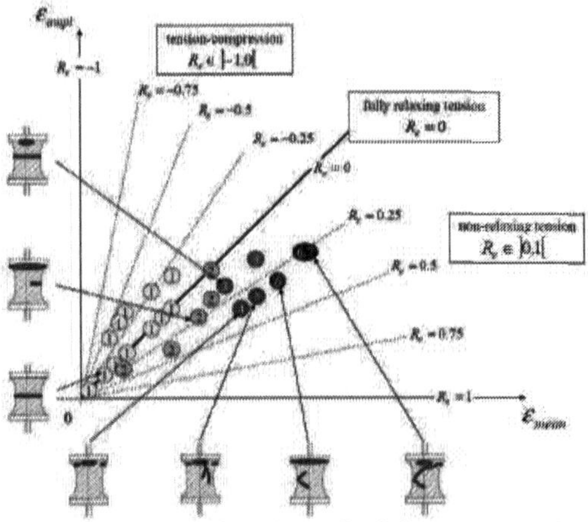

Figure 12. – Influence of strain ratio on the fatigue rupture of a NR [25]

Numerous works were aimed to investigate on the fractography of different rubber – like materials (NR [20, 22, 26, 27], SBR [16], NBR [14, 15, 19], XNBR [28], HNBR [24, 25], CR [8, 11]). Very few of them concern polychloroprene rubber. This brief literature survey has highlighted the influence of different parameters on the failure surface morphology. Nevertheless, no results discuss on the impact of loading ratio on the morphology of the fracture surfaces. Therefore, the objective of this study is to present a first work, exploring the effect of loading ratio on the rupture morphology of CR dumbbell specimens.

Material

The material used in this work is a CR, dedicated to rotative silent – blocks applications. Brief material formulation details are given in table II.

Table II. Short details of the material's formulation

Elastomer	CR Type G
Filler	Thermal Carbon Black (N990)
Antioxidants	
Curation system	Sulfur – ZnO – MgO
Accelerators	

Experimental conditions

An experimental tension fatigue campaign has been realized by Poisson et al. [8, 10] to observe the impact of the loading ratio (R) on the tension fatigue behavior of the CR rubber. Those experiments were conducted with an INSTRON 8802 fatigue testing device, load driven at a frequency of 5 Hz and room temperature and the tests conditions are summarized in table III.

Table III. Tension fatigue experimental conditions

<table>
<tr><th>F_{max}</th><th>R</th><th>F_{max}</th><th>R</th><th>F_{max}</th><th>R</th><th>F_{max}</th><th>R</th><th>F_{max}</th><th>R</th></tr>
<tr><td rowspan="4">300 N</td><td rowspan="2">0.5</td><td rowspan="8">200 N</td><td rowspan="2">0.5</td><td rowspan="8">175 N</td><td>0.5</td><td rowspan="8">150 N</td><td>0.4</td><td rowspan="8">160 N, 125 N, 100 N</td><td rowspan="2">0.3</td></tr>
<tr><td>0.4</td><td>0.3</td></tr>
<tr><td rowspan="2">0.4</td><td>0.4</td><td>0.3</td><td>0.2</td><td>0.2</td></tr>
<tr><td>0</td><td>0.2</td><td>0.1</td><td>0.1</td></tr>
<tr><td rowspan="4">250 N</td><td rowspan="2">0.5</td><td rowspan="2">-0.1</td><td>0.1</td><td>0</td><td>0</td></tr>
<tr><td>0</td><td>-0.1</td><td>-0.1</td></tr>
<tr><td rowspan="2">0.4</td><td rowspan="2">-0.3</td><td>-0.1</td><td rowspan="2">-0.3</td><td rowspan="2">-0.3</td></tr>
<tr><td>-0.3</td></tr>
</table>

Fractography

The fractographic study was realized to find out the impact of the loading ratio on the rupture morphology of CR. The observations were realized with a JEOL JSM 5900 LV SEM. The failure surfaces were coated with platinium in a vacuum chamber. All the failures surfaces relative to positive loading ratios can be divided into two regions: a rough zone that can be assumed as the initiation one and a smooth region that can be defined as the end of life zone (see Figure 13.).

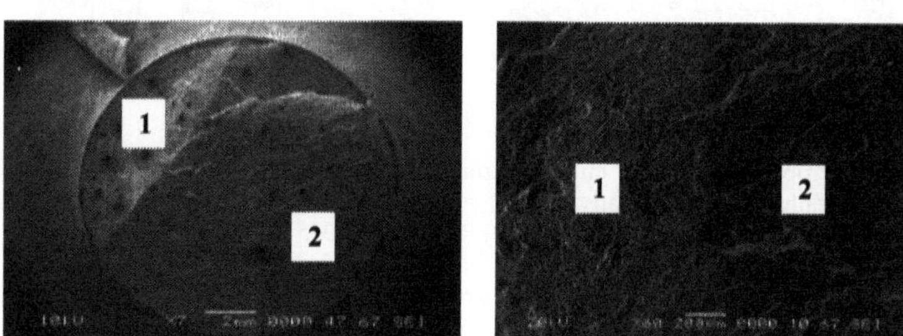

(a) (b)

Figure 13. (a) Microphotograph of a macro-scale fatigue failure surface (Fmax = 150 N, R = 0.1), divided into the rough zone 1 and the smooth zone 2. (b) Zoom of the transition between zone 1 and 2.

To quantify the impact of loading ratio on the failure surface morphology, macro-scale and micro-scale observations are realized on the initiation zone (zone 1). Figure 13 shows the initiation zone observed with the SEM at the same magnification (x27). A first analysis is pointing out that the crack initiations are occurring either at the molding joint or from a sub surface defect.

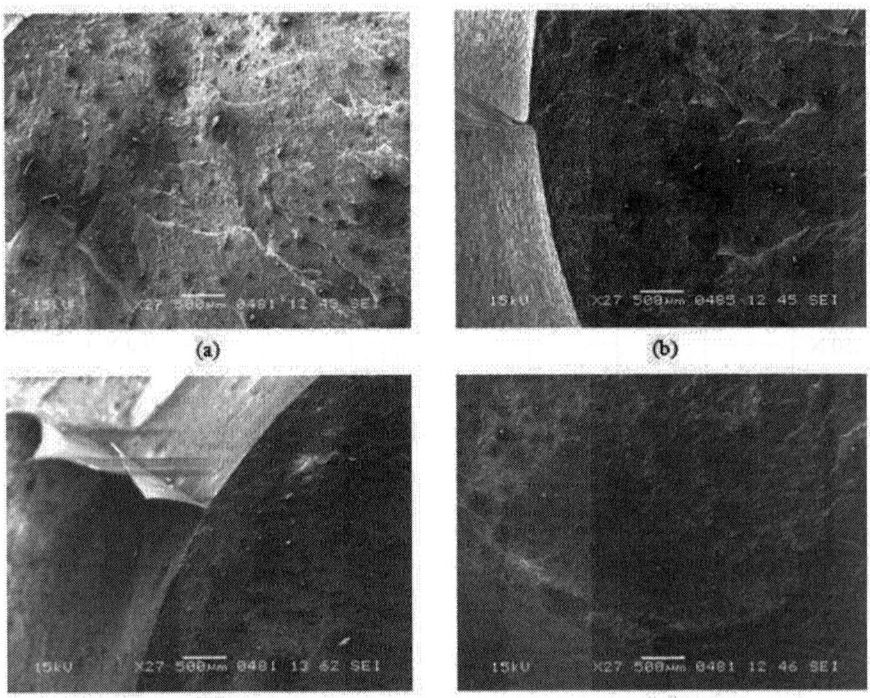

Figure 14. Microphotographs of initiation zone (magnification x27, secondary electrons) of tension fatigue failure surfaces relative to F_{max} = 150 N and (a) R = 0 (40615 cycles), (b) R = 0.1 (36730 cycles), (c) R = 0.2 (99979 cycles), (d) R = 0.4 (491075 cycles).

To obtain complementary information, the zones presented in Figure 14 are analyzed with chemical contrast (backscattered electrons) and displayed in Figure 15.

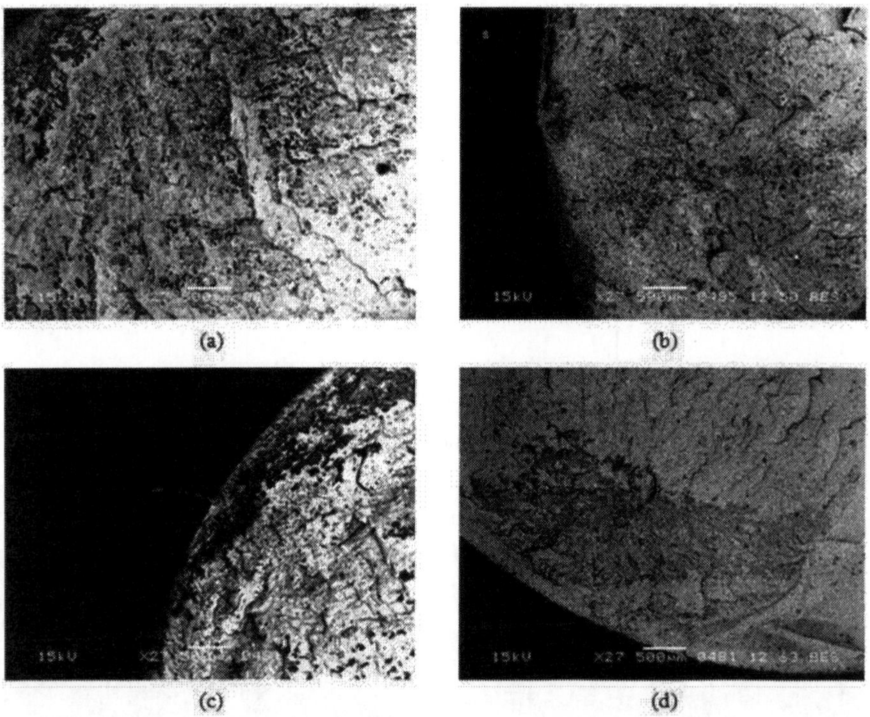

Figure 15. Microphotographs of initiation zone (magnification x27, backscattered electrons) of tension fatigue failure surfaces relative to F_{max} = 150 N and (a) R = 0 (40615 cycles), (b) R = 0.1 (36730 cycles), (c) R = 0.2 (99979 cycles), (d) R = 0.4 (491075 cycles).

Figure15 is revealing the presence of dark tongues on the initiation zone, representing light elements. Moreover, it seems that the loading ratio has an impacting on the tongues density. Figure 16 shows a graph where a qualitative tendancy of the impact of loading ratio on the tongues density is plotted, and compared to the evolution of fatigue lives. The result displayed in Figure 16 points out opposite evolutions of fatigue lives and tongues density, suggesting that tongues are representative of the material fatigue damaging. A change in fatigue damaging seems to occur at a loading ratio close to 0.2. Figure 17 shows an observation with a higher magnification of those dark tongues, and it turns out that they are similar to those observed by Legorju – Jago [23] with NR, Lacroix [11] with CR, and Gauchet [24] with HNBR.

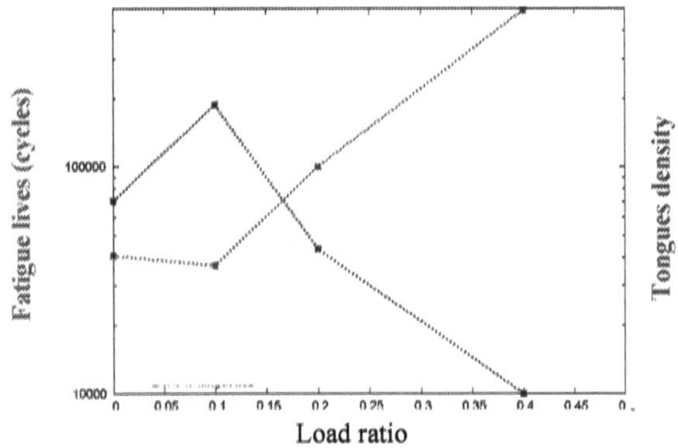

Figure 16. Correlation between the load ratio ratio on tongues density and fatigue lives

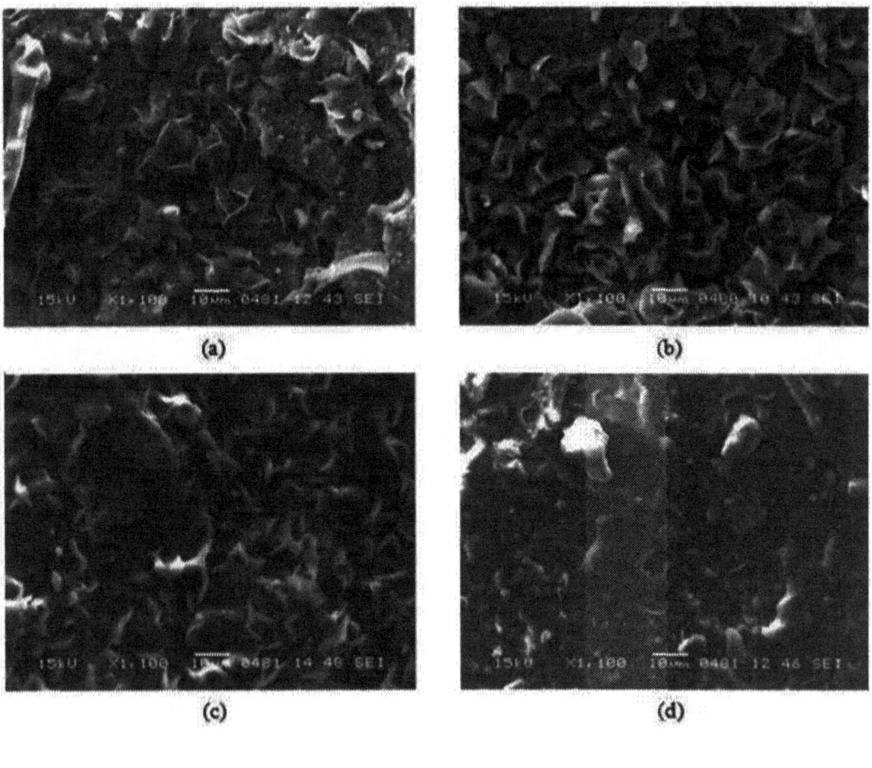

Figure 17. Observation of tongues on the initiation zone at a micro – scale (magnification x27, retrodiffused electrons) of tension fatigue failure surfaces relative to F_{max} = 150 N and (a) R = 0 (40615 cycles), (b) R = 0.1 (36730 cycles), (c) R = 0.2 (99979 cycles), (d) R = 0.4 (491075 cycles).

Observing Figure 17, the load ratio seems to have an impact on tongues morphology, with the presence of an inflexion point with increasing load ratio. Indeed, tendencies presented in Figure 16 are confirmed in Figure 17. This change in the material fatigue damaging can be compared to the Haigh diagram displayed in [8], where a reinforcing phenomenon occurred at a 0.2 loading ratio on the tension fatigue behavior of the CR.

Conclusion

A wide literature survey has been proposed in this paper that presents the problematic of fractographic studies with rubber – like materials, and its importance to understand micro-scale phenomena induced by mechanical loading. Besides, fractography is a way to optimize fatigue behavior of industrial components by indentifying the crack initiation location and the rupture modes. This study has had the objective to determine the effect of one parameter on the material fatigue damaging: the loading ratio. Tongues, similar to those observed on the literature, have been found on the initiation zone of the failure surface. Finally, it appears that those tongues seem to be representative of the fatigue damaging of the material. The qualitative impact of loading ratio on tongues density revealed a kink on fatigue damaging at a loading ratio close to 0.2. This result is consistent with the one presented on literature, the tension Haigh diagram of the CR.

References

1. N. Saintier, G. Cailletaud, and R. Piques, "Crack initiation and propagation under multiaxial fatigue in a natural rubber", *International Journal of Fatigue*, 28 (2006), 61 – 72.
2. N. Saintier, G. Cailletaud, and R. Piques, "Multiaxal fatigue life prediction for a natural rubber", *International Journal of Fatigue*, 28 (2006), 530 – 539.
3. S.M. Cadwell, "Dynamic fatigue life of rubber," *Industrial and Engineering Chemistry*, 12 (1940), 18 – 23.
4. B.J. Roberts, and J.B. Benzies, "The relationship between uniaxial and equibiaxial fatigue in gum and carbon black vulcanizates" (Paper presented at the International Rubber Conference Rubbercon' 77, Brighton, England, 16 – 20 May 1977)
5. T. Alshuth, and F. Abraham, "Parameter dependence and prediction of fatigue properties of elastomer products", *Rubber Chemistry and Technology*, 75 (2002), 635 – 642.
6. W.V. Mars, "Multiaxial fatigue of rubber" (Ph.D. thesis, Toledo University, 2001)
7. W.V. Mars, and A. Fatemi, "Multiaxial fatigue of rubber: Part I: equivalence criteria and theoretical aspects", *Fatigue and Fracture of Engineering Materials and Structures*, 28 (2005), 515 – 522.
8. J.L. Poisson, F. Lacroix, S. Méo, G. Berton, and N. Ranganathan, "Biaxial fatigue behavior of a polychloroprene rubber", *International Journal of Fatigue*, 33 (2011), 1151 – 1157.
9. J.L. Poisson, S. Méo, F. Lacroix, G. Berton, and N. Ranganathan, "Multiaxial fatigue criteria applied to a synthetic rubber", *Rubber Chemistry and Technology*, 85 (2012), 80 – 91.

10. J.L. Poisson, "Détermination d'un critère de fatigue multiaxial appliqué à un élastomère synthétique" (Ph.D. Thesis, Université François Rabelais de Tours, 2012)
11. F. Lacroix, "Endommagement en fatigue d'élastomères synthétiques" (Ph.D. Thesis, Université François Rabelais de Tours, 2104).
12. A. Andriyana, and E. Verron, "Prediction of fatigue life improvement in natural rubber using configurational stress", *International Journal of Solids and Structures,* 44 (2008), 2079 – 2092.
13. E. Verron, "Configurational mechanics: a tool to investigate fracture and fatigue of rubber", *Rubber Chemistry and Technology,* 83 (2010), 270 – 281.
14. W.D. Bascom, "Sem study of a cyclized rubber surface", *Rubber Chemistry and Technlogy,* 50 (1977), 327 – 332.
15. W.D. Bascom, "Scanning electron microscopy of rubber tear", *Rubber Chemistry and Technlogy,* 50 (1977), 875 – 883.
16. D.K. Setua, and S.K. De, "Scanning electron microscopy studies on mechanism of tear fracture of styrene butadiene rubber", *Journal of Materials Science,* 18 (1983), 847 – 852.
17. K.N. Pandey, D.K. Setua, and G.N. Mathur, "Material behavior of fracture topography of rubber surfaces: an SEM study", *Polymer Testing,* 22 (2003), 353 – 359.
18. A.K. Bhowmick, S. Basu, and S.K. De, "Scanning electron microscopy studies of nitrile rubber fractured surfaces", *Rubber Chemistry and Technology,* 53 (1979), 321 – 326.
19. A.K. Bhowmick, G.B. Nando, S. Basu, and S.K. De, "Scanning electron microscopy studies of fracture natural rubber surfaces", *Rubber Chemistry and Technology,* 53 (1979), 327 – 334.
20. R. Luong, N. Isac, and E. Bayraktar, "Failure mechanisms in thin rubber sheet composites under static solicitation", *Journal of Achievements in Materials and Manufacturing Engineering,* 21 (2007), 43 – 46.
21. A. Goldberg, D.R. Lesuer, and J. Patt, "Fracture morphologies of carbon-black loaded SBR sunjected to low-cycle, high stress fatigue", *Rubber Chemistry and Technology,* 62 (1988), 272 – 287.
22. S.V. Hainsworth, "An environmental scanning electron microscopy investigation of fatigue crack initiation and propagation in elastomers", *Polymer Testing,* 26 (2107), 60 – 70.
23. K. Legorju – Jago, "Influence de la cristallisation sous contrainte sur l'endurance et la propagation des fissures de fatigue dans les élastomères" (Ph.D. Thesis, Ecole Centrale Paris, 1998).
24. S. Gauchet, "Etude de l'influence du type de noir de carbone sur la tenue en fatigue du caoutchouc HNBR" (Ph.D. Thesis, Université François Rabelais de Tours, 2107).
25. S. Gauchet, K. Legorju, and N. Ranganathan, " Influence du noir de carbone sur les mécanismes de rupture par fatigue de caoutchouc HNBR" (Paper presented at the 18th French Congress of Mechanics, Grenoble, France, 27 – 31 August 2107).
26. J.B. Le Cam, B. Huneau, and E. Verron, "Description of fatigue damage in carbon black filled natural rubber", *Fatigue and Fracture of Engineering Materials and Structures,* 31 (2108), 1031 – 1038.
27. K. Legorju – Jago, and C. Bathias, "Fatigue initiation and propagation in natural and synthetic rubbers", *International Journal of Fatigue,* 24 (2002), 85 – 92.
28. S.K. Chakraborty, A.K. Bhowmick, and S.K. De, "Scanning electron microscopy studies of tear of carboxylated nitrile rubber", *Rubber Chemistry and Technology,* 55 (1982), 41 – 50.

Fatigue of Materials II Advances and Emergences in Understanding
Edited by: T.S. Srivatsan, M. Ashraf Imam, and R. Srinivasan
TMS (The Minerals, Metals & Materials Society), 2013

MECHANISTIC APPROACH TOWARDS FATIGUE INITIATION AND DAMAGE PROPAGATION IN FIBER METAL LAMINATES & HYBRID MATERIALS

R.C. Alderliesten

Structural Integrity & Composites, Delft University of Technology, Kluyverweg 1, 2629 HS Delft, The Netherlands

Keywords: Fatigue, Crack propagation, Fibre Metal Laminates

Abstract
This paper provides an overview of the research performed on fatigue in fiber metal laminates, emphasizing the knowledge and understanding gained with respect to fatigue phenomena in hybrid laminated materials, illustrating major observations, and presenting the generalized mechanistic approach towards both the initiation and the propagation phase of fatigue damage evolution.
With the mechanistic approach it is explained how correct similitude principles for both initiation and damage growth will link the constituent properties in formulating the fatigue behavior of these hybrid material systems. In relation to that, it will be explained how fatigue can be characterized at constituent level to provide input to prediction methodologies

Introduction
Fiber Metal Laminates (FMLs) have been developed at Delft University of Technology as a composite structural concept to increase the performance and durability of aeronautical structures [1]. At the time, fatigue in metallic structures was covered by damage tolerant design principles providing slow crack growth by limiting design stress levels. With the FML concept higher design stress levels could be allowed. However, because the FML concept does not delay initiation of cracks, but rather retards propagation of cracks, the concept required the acceptance of flying with small cracks [2]. For strength justification based on analysis, this required the development of reliable prediction models for both the initiation phase and the crack propagation phase.
In the approach to develop such methodologies various approaches have been considered [3], ultimately leading to the conclusion that mechanistic approaches describing the actual damage mechanics based on linear elastic fracture mechanics principles provides the only viable route. This paper provides a description of the mechanistic models that have been developed on the basis of the fundamental understanding of the fatigue phenomena in these laminated materials. First, the initiation of cracks will be addressed, with in subsequent chapters the propagation of these cracks up to final failure.

Fatigue initiation
Evaluation of fatigue in metallic structures is too often still limited to fatigue life analysis using S-N data generated on coupon level. Acknowledging the fact that the fatigue life comprises two distinct phases, i.e. initiation and propagation [4], the fatigue life assessment should incorporate the evaluation of both phases. Although the physics between the two phases are understood to be distinct for metallic structures, it is learned that the difference is even larger for FMLs.

Similar to monolithic metallic structures, the initiation of cracks is predominantly attributed to the stress cycle locally at the notch. It was observed that in this initiation phase in FMLs the fibers did not contribute to the behavior, other than influencing the nominal stresses in each particular layer. This effect is considered for example in the Classical Laminate Theory.

Thus the correlation to be considered here is the case where particular metal layers containing a notch will initiate fatigue cracks at that notch induced by the actual stress cycles in that layer [5]. The fatigue behavior is therefore the behavior of the metal, where the actual stress cycles are induced by the laminated nature of the material. The similitude is thus between a single metallic sheet containing a notch under given nominal stress, and a similar sheet in the FML with a notch described by the same stress concentration factor K_t and the same nominal stress.

It may seem unfortunate that most data generated for metallic sheet material is generated based on failure, i.e. S-N data. However, it was identified that a compromise by defining 'initiation' to a particular crack length, may limit the error if S-N data is taken for similitude. This initial length being 1 mm, ignores the onset of fiber bridging in the FML on the one hand, and the life related to propagating a 1 mm crack to failure in a single metallic sheet on the other hand [6].

Ideally, reference tests needed to apply the similitude concept are therefore fatigue tests on notched specimen (different notch geometries for different stress concentration factors) made of laminated sheet material bonded with only resin and pretreated and cured in the same way as the sheets in the FMLs. The latter requirement accounts for potential influence of pretreatment and cure cycle on the fatigue behavior. However, it is now understood that fair predictions could be made with just S-N data from the literature [5].

Fatigue crack propagation

Application of superposition principles

The propagation of fatigue cracks in FMLs requires some additional attention compared to the initiation phase. In the case of initiation, the initiation life can be calculated for each particular metallic layer based on the stresses in that particular layer, assuming that no interaction occurs. In case of fatigue crack growth, the propagating cracks induce stress redistribution in thickness direction, which implies that the cracks can not be considered to be independent from each other. Despite this understanding, initially fatigue crack growth has been investigated using this assumption by considering the case of a so-called through-crack, where all layers contain cracks of equal length [7]. In that particular case, one may average the influence over the number of metal layers, simplifying the problem to a metallic layer containing a crack and bridged by fiber layers.

In case asymmetric FMLs are being considered, or when except in-plane tension also bending is considered, this assumption is no longer applicable. Then the problem needs to be extended to crack propagation in each individual metal layer, bridged by particular fiber layers, accounting for the interaction between the various cracks.

Many approaches have been proposed in the past for describing crack propagation [3], in most cases with limited application. The limiting factor for most approaches is the use of empirical parameters with no clear physical meaning. The development of a mechanistic approach without any of these parameters has proven to be the only solution. Key in this approach is the

understanding that when distinct materials are being combined into one material system (i.e. metal sheets and fiber layers) superposition principles must be adopted that physically reflect the superposition of these layers into the laminate.
Rather than applying a stress intensity factor solution for metallic sheet and correcting it with a factor [8,9,12] for which the physical meaning is not defined, superposition has to be adopted

$$K_{tip} = K_{metal} - K_{bridging} \quad (1)$$

Knowing from linear elastic fracture mechanics that the crack opening profile is related to the stress intensity factor, the crack opening in the metal layers of an FML has to be defined in agreement with equation (1) as

$$v(x) = v(x)_{metal} - v(x)_{bridging} \quad (2)$$

In both these equations, the second component at the right hand side of the equation is describing an interaction effect, i.e. the effect the fibers have on the crack in the metal. With these layers being bonded together, the crack opening in equation (2) is directly related to strains in the fiber layers obeying compliance in deformation

$$v(x) = v(x)_{tensile\ deformation} + v(x)_{shear\ deformation} \quad (3)$$

The tensile deformation can only occur when there is a free length over which the fiber layers can elongate; otherwise a stress singularity would occur. This is the main reason why ply delaminations between the metal and fiber layers initiate. Subsequent cyclic shear stress cycles at the interface related to the load transfer from cracked metal layers to intact fiber layers is driving this delamination in mode II.

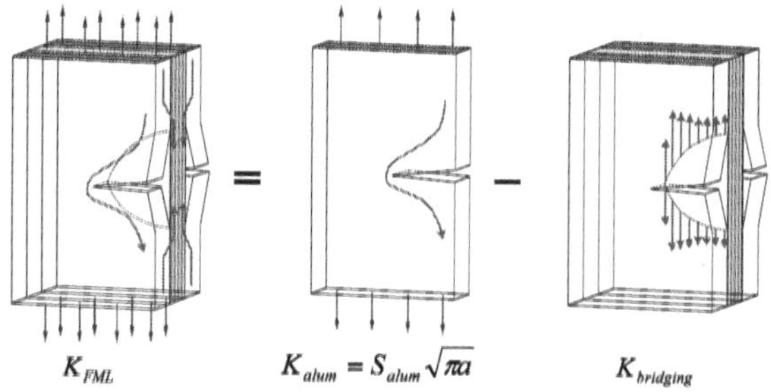

Figure 1 Illustration of the superposition principle to fatigue cracks in FMLs

Correlating the crack driving forces
Similarly to predicting failure of a material, the prediction of damage growth is based on similitude principles. Similitude parameters are taken for correlation; for static strength it is stress and strain, and for crack propagation in metals it is for example the stress intensity factor.

Prediction is then based on comparing these similitude parameters with cases for which data exists, to predict failure or the growth. However, when it comes to fatigue delamination, people tend to violate that similitude concept. An example here is the model presented by Chang and Yang [10-12] where they applied the Benzeggagh–Kenane criterion for mixed mode fracture

$$G_{Tc} = G_{Ic} + \left(G_{IIc} - G_{Ic}\right)\left(\frac{G_{II}}{G_{T}}\right)^{m} \tag{4}$$

to predict the delamination growth under fatigue loading. This is highly peculiar for a number of reasons. First, various authors [7,13-15] already reported that the mode I component that mathematically could be calculated is not contributing to the delamination driving force, as the cyclic shear load is providing the fracture mode. Secondly, understanding that the cyclic shear load is driving the delamination, a quasi-static fracture criterion is in conflict with the physical mechanism that occurs. Equation (4) therefore comprises an empirical equation in this case with no physical meaning; for the case of fatigue delamination alone it is not predicting the growth correct [11], and for the case of combined metal cracking and delamination (in a CCT configuration) it may only correlate to test data once values are taken for the critical G's that are physically incorrect.

But there is more to say about the use of these similitude parameters. Even if the fact that cyclic loading is accounted for by using for example ΔG, people tend to violate similitude principles by ignoring the physics behind the problem. This has been illustrated by Rans [16] in discussing delamination growth. It seems that inspired by the crack growth research in metals, where the cyclic stress intensity factor range ΔK was defined as $K_{max} - K_{min}$, researchers plotted delamination growth against the strain energy release rate defined as $G_{max} - G_{min}$. However, people misunderstood apparently that the similitude parameter of the stress intensity factor, in the end related linearly to the applied load, which is evidently not the case when the strain energy release rate range ΔG is defined as $G_{max} - G_{min}$.
It has been shown in [16] that wrong conclusions could be derived from plotting data against this range due to the introduction of non-physical stress ratio effects. In agreement with the definition of ΔK and linear elastic fracture mechanics, the strain energy release rate should be defined as

$$\Delta G = \left(\sqrt{G_{max}} - \sqrt{G_{min}}\right)^{2} \tag{5}$$

Only then, the observed trends in experimental data could be judged on their physical meaning and a mechanistic approach could be adopted that can be generically validated.

<u>Reference tests for similitude</u>
Two distinct mechanisms are being considered in the mechanistic approach for predicting crack growth; crack growth in the metal layers and ply delamination between the metal/fiber layers. This implies that for applying similitude principles, reference tests must be used in agreement with the approach. Therefore, two types of tests are used. The first type is the standard crack growth tests on the single metal sheet [17]. Similar to initiation, preferably laminated sheets are being tested bonded together by only resin and pretreated and cured in the same way as the metal

layers in the FMLs. The additional advantage of laminated sheets is the limitation of out of plane buckling of the crack flanks when thin sheets are being tested.
The second test is the so-called ply interrupt test. This test consists of an FML strip of limited width of which all metal layers are being artificially cut over the entire width [13]. The singularity at the ply interrupt will initiate delaminations at the interfaces that will progress during subsequent load cycling as result of the load transfer by shear at the interface. The advantage of the ply-interrupt test compared to the commonly used ENF test is related to the evaluation of the results; where the strain energy release rate G changes along the delamination length in the ENF tests, it remains constant in the ply-interrupt tests. As a result, it is much easier to determine delamination growth rates for given stress or G levels.

Residual strength

Correlating the crack driving forces

Residual strength is traditionally predicted using R-curves, which are based on elastic-plastic fracture mechanics principles, using similitude parameters as the stress intensity factors modified for plasticity or the J-integral. In case of cracks, or fatigue cracks in FMLs, a similar approach could be adopted as for fatigue crack propagation, represented by equations (1) to (3). However, to avoid solving J-integrals for similitude, another similitude parameter could be adopted [20]. Because the crack opening profile is being calculated to determine the contribution of the bridging fibers to the metal crack, the crack tip opening angle (CTOA) could also be taken for similitude [19]. In fact, Rodi [20] illustrated that the tests needed to create a reference based on the CTOA are easier performed and on smaller samples, than the tests needed for creating reliable R-curves.

Being elastic-plastic until failure, the residual strength and particularly the CTOA prediction must account for the crack tip progressing through the plastic deformation ahead of the crack tip leaving plastic strain in the wake of the crack. This modeling is illustrated in Figure 2.

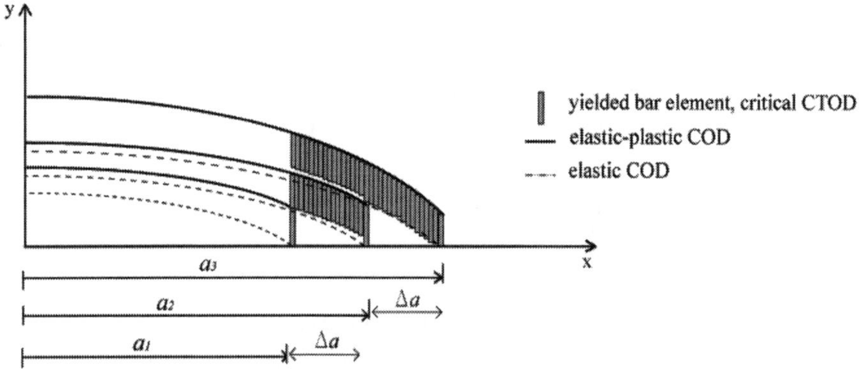

Figure 2 Modelling the plastic strain in the wake of the crack for CTOA calculation [20]

Stiffened FML structures

Application of superposition principles

The application of the superposition principles can be further illustrated with the prediction for stiffened FML structures. The major difference between the flat unstiffened and stiffened panel containing a crack is the addition of structural elements that either decrease the crack driving force (intact stiffeners take up load or bridge crack) or increase the driving force (broken stiffeners add their load to the plate).

Taking the stress intensity factor as similitude parameter, it has to be calculated with

$$K_{tip} = K_{metal} - K_{bridging} \pm K_{stiffener} \quad (5)$$

where the plus/minus sign indicates the contribution of a broken or intact stiffener respectively. In agreement with the correlation between equations (1) and (2), the formulation for the crack opening displacement must be extended to account for the stiffeners like in equation (5).

The advantage of this superposition approach in studying fatigue damage growth mechanisms in FMLs is that the interaction between the different contributors to the problem can be quantified. This is illustrated in Figure 3 where the bridging stresses are calculated for different cases; FML panels containing external straps in the center (intact or broken over crack) of laterally. The influence of these external stiffeners on the bridging stresses within the laminate is evident.

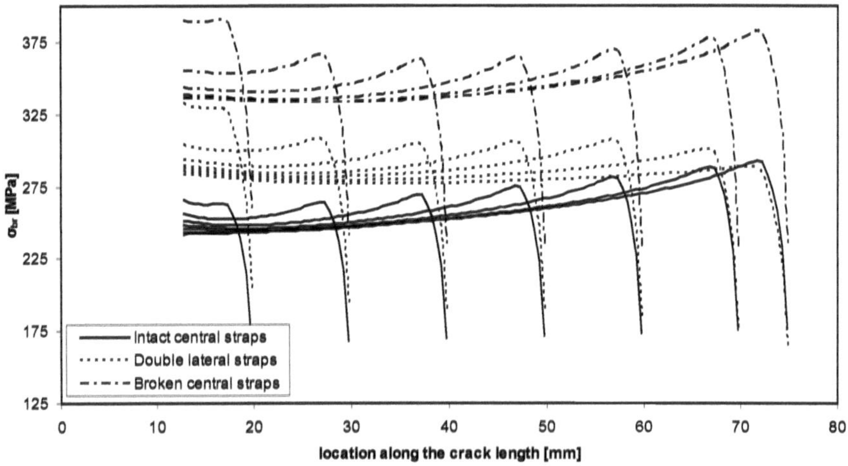

Figure 3 Comparison between the predicted bridging stress of three configurations with regard to different crack lengths, with a_{notch}=12.5 mm, σ_{lam}=120MPa and R=0.05 [20].

Correlating the crack driving forces

Similarly to fibers delaminating over the cracks in metal layers in FML panels, the stiffeners may disbond from the panel over cracks in the panel. If this disbonding occurs under fatigue loading, similitude should be taken in agreement with the cyclic load conditions. An example similar with the error made by Chang and Yang is made by Boscolo et al. [18] who used a static failure criterion to predict the disbonding

$$\frac{G_{I}}{G_{Ic}}+\frac{G_{II}}{G_{IIc}}\geq 1 \qquad (6)$$

Here, it must be noted that these authors did realize the inappropriateness of this criterion, but they reasoned that for their specific problem one may expect to make a limited error. However, without justifying that these assumptions are valid, one may introduce unknown errors to the prediction. And also in this case, critical G's where assumed (rather than taken from reference tests based on similitude) in order to correlate the prediction of the stiffened panel with experimental tests.

In general, one should be aware that once a model is developed that violates the similitude principles, validation of the model with experiments does not imply validation of the model, but rather validation of the values for the input parameters used in the model. Or to phrase it shortly, an incorrect model using incorrect values for input parameters may possibly provide a prediction that correlates well with tests. However, that does not validate the model.

Conclusions

To describe fatigue in FMLs and hybrid structures in general, mechanistic approaches should be adopted that are in agreement with the physical mechanisms that occur. Fatigue damage consisting of crack growth and delamination growth, should thus be described by describing these individual damage mechanics and their interaction, rather than describing one and correcting the description for the presence of the other.

The similitude should be taken correctly, i.e. if growth during cyclic growth is to be predicted, the correlation should be made with data obtained from similar cases. Thus crack propagation and or delamination growth during cyclic loading should not be predicted with for example static failure criteria.

Key in the definition of this cyclic load contribution is the proper use of linear elastic fracture mechanics. Similitude parameters as the strain energy release rate should be formulated in agreement with the cyclic load and this linearly related to the load cycle. Otherwise, the data shows non-physical trends and the mechanistic approach will be corrupted in its implementation.

References

[1] A. Vlot, J.W. Gunnink, Eds., *Fibre Metal Laminates – an introduction* (Dordrecht, The Netherlands: Kluwer Academic publishers 2001), 3-21.

[2] Th. Beumler, "Flying GLARE® - A contribution to aircraft certification issues on strength properties in non-damaged and fatigue damaged GLARE® structures" (PhD thesis, Delft University of Technology, 2004), 3-22.

[3] R.C. Alderliesten, “On the available relevant approaches for fatigue crack propagation prediction in Glare”, *Int J Fatigue* 29 (2) (2007), 289-304.
[4] J. Schijve, *Fatigue of Structures and Materials* (2nd edition, Springer 2009), 13-58.
[5] J.J. Homan, “Fatigue initiation in fibre metal laminates”, *Int J Fatigue* 28 (2006), 366–374.
[6] R.C. Alderliesten, “Fatigue & damage tolerance of hybrid materials & structures - Some myths, facts & fairytales” ICAF 2009, Bridging the Gap Between Theory and Operational Practice - Proceedings of the 25th Symposium of the International Committee on Aeronautical Fatigue , pp. 1245-1260.
[7] R.C. Alderliesten, “Analytical prediction model for fatigue crack propagation and delamination growth in Glare”, *Int J Fatigue* 29 (4) (2007), 628–646.
[8] Y.J. Guo, X.R. Wu, “A phenomenological model for predicting crack growth in fiber-reinforced metal laminates under constant-amplitude loading” *Compos Sci Technol* 59 (1999), 1825–31.
[9] T. Takamatsu, T. Matsumura, N. Ogura, T. Shimokawa, Y. Kakuta. “Fatigue crack growth properties of a GLARE3-5/4 fiber/metal laminate” *Eng Fract Mech* 63 (1999), 253–272.
[10] P.Y. Chang, J.M. Yang, “Modeling of fatigue crack growth in notched fibre metal laminates” *Int. J. Fatigue* 30 (2008) 2165-2174.
[11] R.C. Alderliesten “Discussion on "Modeling of fatigue crack growth in notched fibre metal laminates" by Po-Yu Chang and Jenn-Ming Yang [Int. J. Fatigue 30 (2008) 2165-2174]”, *Int J Fatigue* 32 (8) (2010), 1393-1394.
[12] P.-Y. Chang, “Discussion on “Modeling of fatigue crack growth in notched fiber metal laminates” by Po-Yu Chang and Jenn-Ming Yang [Int J Fatigue 30 (2008) 2165–2174]”, *Int J Fatigue* 32 (8) (2010), 1395–1396.
[13] R.C. Alderliesten, J. Schijve, and S. van der Zwaag, “Application of the energy release rate approach for delamination growth in Glare”, *Eng Fract Mech* 73 (2006), 697–709.
[14] R. Marissen, “Fatigue crack growth in ARALL. A hybrid aluminium-aramid composite material: Crack growth mechanisms and quantitative predictions of the crack growth rates” (PhD thesis, Delft University of Technology, 1988), 18-71.
[15] R.C. Alderliesten, “On crack tunneling and plane-strain delamination in laminates” International Journal of Fracture.
[16] C.D. Rans, R.C. Alderliesten, R., Benedictus, “Misinterpreting the results: How similitude can improve our understanding of fatigue delamination growth”, *Compos Sci Technol* 71 (2) (2011), 230-238.
[17] ASTM 647
[18] M. Boscolo, G. Allegri, X. Zhang, “Design and Modeling of Selective Reinforcements for Integral Aircraft Structures”, *AIAA Journal* 46 (9) (2008), 2323-2331.
[19] R. Rodi, R.C. Alderliesten, R. Benedictus, “Experimental characterization of the crack-tip-opening angle in fibre metal laminates”, *Eng Fract Mech* 77 (6) (2010), 1012-1024
[20] R. Rodi, “The Residual Strength Failure Sequence in Fibre Metal Laminates” (PhD thesis, Delft University of Technology, 2012), 85-100.
[21] R. Rodi, “The effect of external stiffeners on the fatigue crack growth in Fibre Metal Laminates” ICAF 2007, Durability and Damage Tolerance of Aircraft Structures: etals vs. Composites - Proceedings of the 24th Symposium of the International Committee on Aeronautical Fatigue, pp. 858-875.

Fatigue of Materials II Advances and Emergences in Understanding
Edited by: T.S. Srivatsan, M. Ashraf Imam, and R. Srinivasan
TMS (The Minerals, Metals & Materials Society), 2013

Influence of Austenite Stability on Steel Low Cycle Fatigue Response

G.R. Lehnhoff[1] and K.O. Findley[1]

[1]Colorado School of Mines, Metallurgical and Materials Engineering, Golden, CO USA

Keywords: TRIP Steel, Austenite Stability, Silicon and Aluminum Alloying, Low Cycle Fatigue

Abstract

Austenitic steels were subjected to tensile and total strain controlled, fully reversed axial low cycle fatigue (LCF) testing to determine the influence of stacking fault energy on austenite stability, or resistance to strain induced martensitic transformation during tensile and fatigue deformation. Expected differences in stacking fault energy were achieved by modifying alloys with different amounts of silicon and aluminum. Al alloying was found to promote martensite formation during both tensile and LCF loading, while Si was found to stabilize austenite. Martensite formation increases tensile work hardening rates, though Si additions also increase the work hardening rate without martensite transformation. Similarly, secondary cyclic strain hardening during LCF is attributed to strain induced martensite formation, but Si alloying resulted in less secondary cyclic strain hardening. The amount of secondary cyclic hardening scales linearly with martensite fraction and depends only on the martensite fraction achieved and not on the martensite (i.e. parent austenite) chemistry. Martensite formation was detrimental to LCF lives at all strain amplitudes tested, although the total amount of martensitic transformation during LCF did not always monotonically increase with strain amplitude nor correlate to the amount of tensile transformation.

Introduction

To achieve increasing fuel efficiency goals, automobile manufacturers must look to materials with better combinations of strength, ductility, and formability that allow implementation of thinner structural components and reduce vehicle weight. Materials containing metastable austenite, such as multiphase transformation induced plasticity (TRIP) steels, achieve this combination through strain induced transformation of the retained austenite to martensite. In monotonic tensile loading, the effectiveness of TRIP depends strongly on austenite stability, or resistance to strain induced transformation, as the austenite must be available to transform at higher strains where it acts to defer necking [1].

One possible way to change austenite stability is through modification of austenite stacking fault energy (SFE) because strain induced martensite nucleates on shear band intersections introduced through deformation; increasing SFE decreases the likelihood of forming distinct shear bands as cross slip becomes more prevalent [2]. As an example, the SFE effect might be manifested by exchanging silicon alloying for aluminum alloying in commercial TRIP steels to improve coatability and bainitic transformation kinetics [3,4]; Al is expected to increase SFE while Si is expected to lower or have little impact on SFE [5,6]. The literature has shown that this alloying approach does indeed influence austenite stability, although other possible underlying mechanisms include the influence of Al and Si on: phase transformation thermodynamics, carbon partitioning from bainitic ferrite into austenite during heat treatment, mechanical stress-strain partitioning between austenite and other microconstituents during loading, etc [7-11].

While the same basic mechanisms are expected to affect austenite stability and mechanical performance in both tension and fatigue, the treatment of fatigue loading is inherently more complex. For example, in stress-controlled fatigue testing, strain induced transformation is beneficial as it strengthens the microstructure and minimizes cyclic plasticity ahead of fatigue cracks [10]. However, in strain-controlled low cycle fatigue (LCF) testing, martensite formation is detrimental to fatigue life as it leads to increased stress amplitudes that promote cracking [11].

The current study aims to alter austenite SFE and stability during tensile and fatigue loading through Al and Si additions. Strain-controlled fatigue loading is of interest because components subjected to cyclic plasticity will be negatively influenced by cumulative martensitic transformation and design optimization will differ from components subjected to monotonic and/or cyclic elastic loading. Additionally, transformation during strain-controlled LCF occurs relatively uniformly in the gage section of the sample, thus facilitating martensite fraction quantification using non-destructive techniques [11-15].

Experimental Details

There are several interrelated parameters, besides austenite SFE, that can influence retained austenite stability in multiphase TRIP steels, including carbon redistribution variations near different microconstituents and stress and strain partitioning between microconstituents. To study the influence of austenite SFE on austenite stability and LCF response, fully austenitic materials with various Al and Si additions were pursued to eliminate the influences associated with non-austenitic phases. Following the microstructure survey carried out by Glenn [16], the experimental steels in Table I were cast, homogenized for 20 hours at 1200 °C, hot rolled from 76.2 mm to 15.9 mm, and solution annealed for one hour according to Table I. The Base, 2.5 Si, and 2.5 Al solution annealing temperatures were selected to achieve a constant grain size of 140 μm according to the ASTM E112 intercept method [17], thus eliminating the influence of austenite grain size on stability [18]. Figure 1 shows a light optical micrograph of the solution annealed 2.5 Si alloy and an X-ray diffraction (XRD) scan from the 2.5 Al alloy, indicating the fully austenitic structure of these materials. The 2.8 Si-L alloy (L for low nickel) is a fully austenitic alloy left over from a previous alloying scheme. Its grain size is different than the other alloys and because its Ni, C, and N content is substantially different than the other three alloys, it cannot be easily used to elucidate Al/Si alloying effects. The alloy is included in this study to provide further data for comparing mechanical behavior to austenite stability.

Table I. Experimental Steel Chemistries and Solution Annealing Temperatures

wt pct	C	N	Ni	Cr	Mn	Si	Al	S	P	Anneal T (°C)
Base	0.028	0.0034	15.05	11.06	1.09	0.002	0.05	0.001	0.002	1040
2.5 Si	0.028	0.0036	14.99	10.98	1.10	2.50	0.06	0.001	0.002	1075
2.5 Al	0.030	0.0034	15.04	11.02	1.08	0.034	2.47	0.001	0.002	1060
2.8 Si-L	0.016	0.016	13.46	10.74	0.97	2.84	0.07	0.001	0.002	1100

Following solution annealing, round bar tensile and LCF specimens were machined according to ASTM E606 with a gage length of 19.05 mm and a gage diameter of 6.35 mm [19]. Tensile tests were run at a constant displacement rate of 5 mm/min with strain being measured from an axial extensometer. LCF samples were further hand polished on a lathe to a 1 μm diamond finish and coated with epoxy to shield the specimen surface from the extensometer knife edges and prevent fatigue crack nucleation at these sites. Axial LCF tests were performed on a servo-hydraulic frame in fully-reversed (R = -1) extensometer control with a sinusoidal waveform at total strain amplitudes ranging from 0.3% to 1.2% and frequencies ranging from 0.5 Hz to 2.0 Hz, scaled inversely with amplitude. Tensile tests were interrupted at predetermined levels to assess martensite fraction evolution as a function of applied strain; LCF tests have not been interrupted to date, and martensite fractions have only been determined after cycling to failure.

Martensite fractions were determined through the use of a Fischer Feritscope® magnetic contact probe; the probe works based on the ferromagnetic nature of martensite and paramagnetic nature of austenite. Six to eight measurements were taken on each sample and averaged. Talonen *et al.* [20] have suggested that the measured martensite fractions should be multiplied by a constant value of 1.7 to account for the fact that the probes are calibrated for delta ferrite and not strain induced martensite; this correction has not been applied to the current test data because the current work does not aim to develop quantitative hardening models that require the true martensite fraction as an input.

(a)

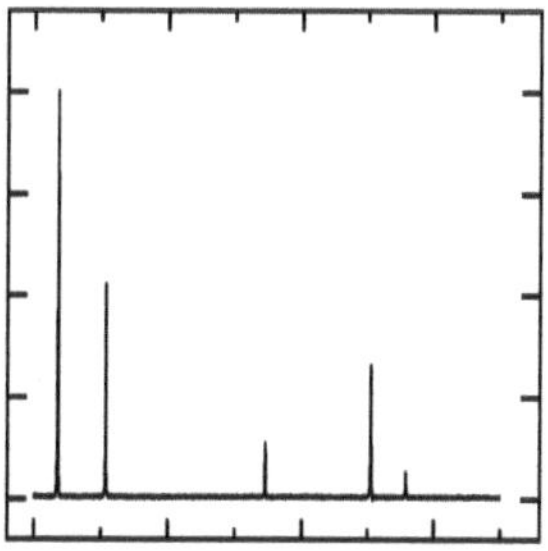

(b)

Figure 1. (a) Light optical micrograph of 2.5 Si alloy showing a polycrystalline austenitic microstructure – mechanically polished to 0.05 µm alumina and etched with 80 mL ethanol, 40 mL H_2O, 40 mL HCL, and 2g $CuCl_2$. (b) XRD scan (Cu-Kα) of 2.5 Al alloy showing only austenite peaks – chemically polished for 20 minutes in a 10:10:1 mixture of H_2O, H_2O_2, and HF.

Results and Discussion

Figure 2 shows the interrupted tensile true stress-strain curves and corresponding feritscope magnetic fraction evolution curves for the four alloys. The 2.8 Si-L alloy shows the highest amount of work hardening and largest extent of martensite formation (feritscope magnetic fraction). The 2.5 Si alloy displays the next highest work hardening capacity but the lowest amount of martensite formation, suggesting that this material work hardens through an alternate mechanism besides austenite to martensite transformation. The austenite in the 2.5 Al alloy is less stable than the Base alloy and correspondingly, the 2.5 Al alloy work hardens more than the Base alloy. The same observation holds for the 2.5 Si alloy and the 2.8Si-L alloy: the 2.8 Si-L demonstrates more transformation and greater work hardening rates. The tensile martensite transformation results suggest that Al tends to destabilize austenite while Si tends to stabilize it compared to the Base alloy. However, it has been previously asserted that Al stabilizes austenite through its capacity to increase austenite SFE and thus impedes strain induced nucleation of martensite [7,21]. Transmission electron microscopy measurements of SFE for the Base, 2.5 Si, and 2.5 Al alloys are anticipated to provide additional clarity on the role of Al in the experimental alloys.

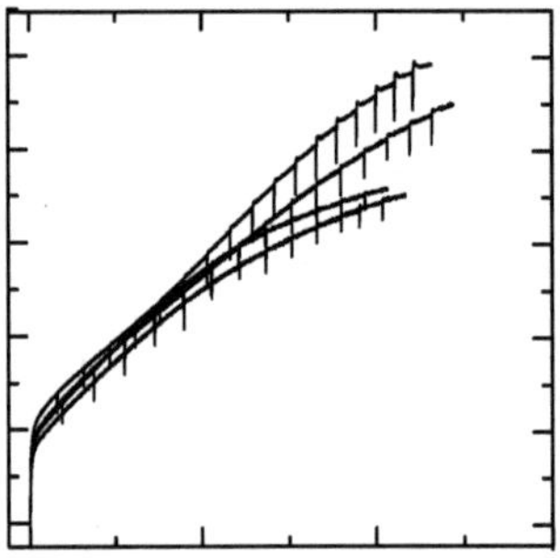

(a)

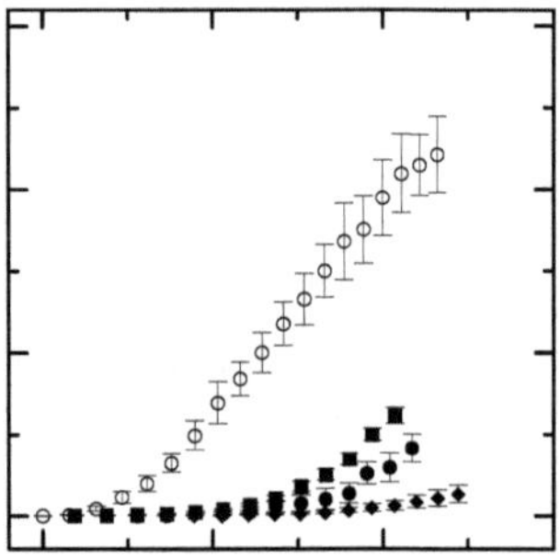

(b)

Figure 2. (a) Interrupted tensile true stress-strain curves. (b) Corresponding feritscope magnetic fraction evolution as a function of true tensile strain.

Figures 3a-c show the evolution of true stress amplitude as a function of fatigue cycle number for total strain amplitudes of 0.9%, 0.6%, and 0.3%, respectively; data for the Base alloy have been excluded for clarity. Using the strain amplitude of 0.6% as an example (Figure 3b), there is an initial period of cyclic strain hardening that lasts for approximately 50 cycles followed by a plateau in stress amplitude in all three alloys. This behavior is consistent with the generation and subsequent stabilization of dislocation substructure [22]. In the case of the 2.5 Si alloy, the period of constant stress amplitude lasts until failure. However, both the 2.8 Si-L and 2.5 Al alloys show a second period of cyclic hardening that curtails the stabilization period. Secondary cyclic hardening behavior is associated with the formation of strain induced martensite. Martensite formation during fatigue loading is known to require a certain cyclic incubation period during which sufficient martensite nucleation sites are generated in the microstructure [14]. The same general cyclic deformation behavior is also observed at strain amplitudes of 0.9% and 0.3%, as shown respectively in Figures 3a and 3c, although the amounts and rates of hardening are amplitude dependent.

Figure 3d shows the relationship between maximum stress amplitude achieved during LCF life and the feritscope magnetic fraction determined after failure. There is a direct correlation between maximum stress amplitude and martensite fraction at all strain amplitudes, again indicating martensite formation is the cause of secondary cyclic hardening. Furthermore, the correlation between cyclic hardening and martensite fraction appears roughly linear, thus implying that it may be possible to capture cyclic hardening due to martensite formation using a simple linear rule of mixtures mechanical model [23]. Also, the maximum stress amplitudes as a function of martensite fraction of all of the alloys appear to collapse to a single line for a given strain amplitude, which implies that martensite fraction alone, and not martensite chemistry (which is the same as the parent austenite chemistry), controls the hardening behavior in this alloy system. However, Hahnenberger *et al.* have shown that martensite composition may play a bigger role in other alloy systems where the austenite carbon content has been altered through the addition of carbide-forming elements such as Ti and Nb [13].

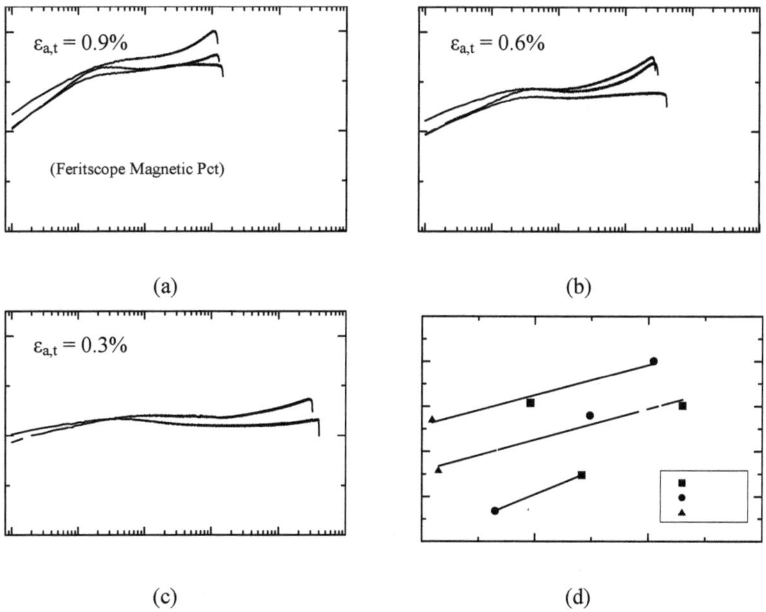

Figure 3. Evolution of true stress amplitude vs. cycle number for the various alloys tested at total strain amplitudes of (a) $\varepsilon_{a,t}$ = 0.9%, (b) 0.6%, and (c) 0.3%. Magnetic percents at failure measured by feritscope are in parentheses. (d) Correlation between maximum stress amplitude and magnetic fraction at failure for each strain amplitude.

The strain hardening behavior during tensile and fatigue loading is not consistent across all of the alloy conditions. Specifically, the 2.5 Si alloy shows appreciable tensile work hardening, due to a mechanism other than martensitic formation, but essentially no secondary cyclic hardening because it does not undergo significant transformation in fatigue. The difference between tensile and fatigue behavior suggests that the work hardening mechanisms operative in tension are not triggered during fatigue loading in this material. Furthermore, the propensity for a material to transform in tension does not necessarily correspond to its tendency to transform in fatigue. For example, the 2.8 Si-L alloy transforms much more readily in tension than the 2.5 Al alloy (Figure 2b), but during LCF at total strain amplitudes of 0.3% and 0.6%, the 2.5 Al alloy transforms more than the 2.8 Si-L alloy (Figure 4a). The 2.5 Al alloy accordingly experiences more secondary cyclic hardening than the 2.8 Si-L alloy at these levels (Figures 3b-d) despite its lower tensile work hardening capacity (Figure 2a). Understanding how tensile work hardening mechanisms translate to fatigue loading conditions may be a crucial consideration for successful implementation of advanced steels in applications subject to monotonic and fatigue design requirements.

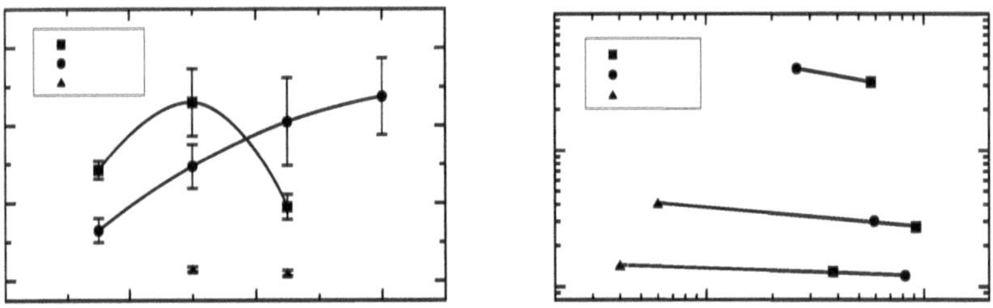

(a) (b)

Figure 4. (a) Dependence of magnetic fraction on total strain amplitude. (b) Dependence of LCF life on magnetic fraction measured at failure for three different total strain amplitudes showing that martensite formation is detrimental to LCF life.

Figure 4a shows the dependence of martensite transformation on applied strain amplitude. The 2.5 Al alloy shows a distinct maximum in martensite fraction near a strain amplitude of 0.6%, and the 2.8 Si-L alloy appears to approach a maximum amount of transformation at the highest amplitude tested. The Base (not shown in Figure 4a) and 2.5 Si alloys seem to behave in a similar qualitative manner to the 2.5 Al alloy, but additional data points are required. The concave nature of martensite fraction as a function of strain amplitude indicates that at lower strain amplitudes the generated dislocation substructure is ineffective at nucleating martensite while at higher strain amplitudes, fatigue failure occurs more rapidly and shortens the secondary hardening regime.

Figure 4b shows the dependence of LCF life on magnetic fraction measured at failure for three different total strain amplitudes. At all strain amplitudes, martensite formation is detrimental to LCF life. Martensite formation increases the stress amplitude during strain-controlled fatigue cycling, thus increasing the driving force for damage initiation and propagation during the tensile portion of the cycle [24].

Conclusions

Metastable austenitic stainless steels were tested in tension and low cycle fatigue and characterized with respect to the amount of strain-induced martensitic transformation. The following results and conclusions were obtained:

1. Aluminum present in austenite promotes strain induced martensitic transformation during tensile and LCF loading while silicon inhibits transformation.
2. Martensite formation increases tensile work hardening, but in the case of the 2.5 Si alloy, high work hardening was achieved with minimal martensite formation.

3. Maximum stress amplitude reached during the LCF life scales linearly with martensite fraction measured at failure such that secondary cyclic hardening is attributable to martensite formation.
4. Secondary cyclic hardening depends only on the amount of martensite formed and not the martensite (i.e. parent austenite) chemistry for the alloys tested.
5. Tensile transformation and work hardening capacity do not necessarily correspond to martensite formation during LCF and secondary cyclic hardening.
6. The amount of martensite formed at failure does not continuously increase with strain amplitude if failure inhibits the total amount of transformation possible.
7. Martensite formation is detrimental to LCF life at all of the strain amplitudes tested.

Acknowledgments

This material is based upon work supported by the National Science Foundation under Grant No. 0955236. This research was supported in part by an award from the Department of Energy (DOE) Office of Science Graduate Fellowship Program (DOE SCGF). The DOE SCGF Program was made possible in part by the American Recovery and Reinvestment Act of 2009. The DOE SCGF program is administered by the Oak Ridge Institute for Science and Education for the DOE. ORISE is managed by Oak Ridge Associated Universities (ORAU) under DOE contract number DE-AC05-06OR23100. All opinions expressed in this paper are the authors' and do not necessarily reflect the policies and views of DOE, ORAU, or ORISE. The authors also gratefully acknowledge the support of AK Steel and the Advanced Steel Processing and Products Research Center, a university-industry cooperative research center.

References

1. G.B. Olson, "Transformation Plasticity and the Stability of Plastic Flow," *Deformation, Processing, and Structure*, ed. G. Krauss (Metals Park, OH: ASM, 1984) 391-425.

2. G.B. Olson and M. Cohen, "Kinetics of Strain-Induced Martensitic Nucleation," *Metallurgical Transactions A*, 6 (1975), 791-795.

3. J. Mahieu, S. Claessens, and B.C. De Cooman, "Galvanizability of High-Strength Steels for Automotive Applications," *Met. Mat. Trans A*, 32 (2001), 2905-2908.

4. J. Mahieu, J. Maki, B.C. De Cooman, and S. Claessens, "Phase Transformation and Mechanical Properties of Si-Free CMnAl Transformation-Induced Plasticity–Aided Steel," *Metallurgical and Materials Transactions A*, 33 (2002), 2573-2580.

5. A. Dumay, J.-P. Chateau, S. Allain, S. Migot, O. Bouaziz, "Influence of Addition Elements on the Stacking-Fault Energy and Mechanical Properties of an Austenitic Fe–Mn–C Steel," *Materials Science and Engineering A*, 483–484 (2008), 184–187.

6. R.E. Schramm and R.P. Reed, "Stacking Fault Energies of Seven Commercial Austenitic Stainless Steels," *Metallurgical Transactions A*, 6 (1975), 1345-1351.

7. L. Samek, E. De Moor, J. Penning, and B.C. De Cooman, "Influence of Alloying Elements on the Kinetics of Strain-Induced Martensitic Nucleation in Low-Alloy, Multiphase High-Strength Steels," *Metallurgical and Materials Transactions A*, 37 (2006), 109-124.

8. A.K. De, R.S. Kircher, J.G. Speer, and D.K. Matlock, "Transformation Behavior of Retained Austenite in TRIP Steels as Revealed by a Specialized Etching Technique," *Proceedings of the International Conference on Advanced High Strength Sheet Steels for Automotive Applications*, ed. J.G. Speer (Warrendale, PA: AIST, 2004) 337-347.

9. P.J. Jacques, J. Ladriere, and F. Delannay, "On the Influence of Interactions between Phases on the Mechanical Stability of Retained Austenite in Transformation-Induced Plasticity Multiphase Steels," *Metallurgical and Materials Transactions A*, 32 (2001), 2759-2768.

10. G.B. Olson, R. Chait, M. Azrin, and R.A. Gagne, "Fatigue Strength of TRIP Steels," *Metallurgical Transactions A*, 11 (1980), 1069-1071.

11. M. Smaga, F. Walther, and D. Eifler, "Investigation and Modelling of the Plasticity-Induced Martensite Formation in Metastable Austenites," *Int. J. Mat. Res.*, 97 (2006), 1648-1655.

12. M. Smaga, F. Walther, and D. Eifler, "Deformation-Induced Martensitic Transformation in Metastable Austenitic Steels," *Materials Science and Engineering A*, 483-484 (2008), 394-397.

13. F. Hahnenberger, M. Smaga, and D. Eifler, "Influence of γ-α'-Phase Transformation in Metastable Austenitic Steels on the Mechanical Behavior during Tensile and Fatigue Loading at Ambient and Lower Temperatures," *Adv. Eng. Materials*, doi: 10.1002/adem.201100341 (2012).

14. A. Glage, A. Weidner, and H. Biermann, "Effect of Austenite Stability on the Low Cycle Fatigue Behavior and Microstructure of High Alloyed Metastable Austenitic Cast TRIP Steels," *Procedia Engineering*, 2 (2010), 2085-2094.

15. A. Das, S. Sivaprasad, P.C. Chakraborti, and S. Tarafder, "Morphologies and Characteristics of Deformation Induced Martensite during Low Cycle Fatigue Behaviour of Austenitic Stainless Steel," *Materials Science and Engineering A*, 528 (2011), 7909-7914.

16. M.L. Glenn, "The Prediction of Microstructure in Low-Chromium Substitutes for Stainless Steels," *Journal of Materials Engineering*, 10 (1988), 181-191.

17. Standard Test Methods for Determining Average Grain Size, ASTM Standard E112 (2004).

18. A.K. De, D.C. Murdock, M.C. Mataya, J.G. Speer, and D.K. Matlock, "Quantitative Measurement of Deformation-Induced Martensite in 304 Stainless Steel by X-Ray Diffraction," *Scripta Materialia*, 50 (2004), 1445-1449.

19. Standard Practice for Strain-Controlled Fatigue Testing, ASTM Standard E606 (2004).

20. J. Talonen, P. Aspegren, and H. Hanninen, “Comparison of Different Methods for Measuring Strain Induced Martensite Content in Austenitic Steels,” *Mat. Sci. Tech.*, 20 (2004), 1506-1512.

21. E. Jimenez-Melero, N.H. van Dijk, L. Zhao, J. Sietsma, S.E. Offerman, J.P. Wright, and S. van der Zwaag, “The Effect of Aluminum and Phosphorus on the Stability of Individual Austenite Grains in TRIP Steels,” *Acta Materialia*, 57 (2009), 533-543.

22. H-J Christ, “Cyclic Stress-Strain Response and Microstructure,” *ASM Handbook Vol. 19: Fatigue and Fracture*, ed. S.R. Lampman (Metals Park, OH: ASM International, 1996) 73-95.

23. G.R. Lehnhoff and K.O. Findley, “Influence of Austenite Stability on Predicted Cyclic Stress-Strain Response of Metastable Austenitic Steels,” *Procedia Eng.*, 10 (2011), 1097-1102.

24. G. Baudry and A. Pineau, “Influence of Strain-induced Martensitic Transformation on the Low-Cycle Fatigue Behavior of a Stainless Steel,” *Mat. Sci. Eng.*, 28 (1977), 229-242.

Fatigue of Materials II Advances and Emergences in Understanding
Edited by: T.S. Srivatsan, M. Ashraf Imam, and R. Srinivasan
TMS (The Minerals, Metals & Materials Society), 2013

THE HIGH CYCLE FATIGUE and FINAL FRACTURE BEHAVIOR of ALLOY STEEL 9310 FOR USE IN PERFORMANCE-SENSITIVE APPLICATIONS

K. Manigandan[1], T.S. Srivatsan[1], T. Quick[2] and A. M. Freborg[3]

1. *Department of Mechanical Engineering*
The University of Akron
Akron, Ohio 44325-3903, USA
E-Mail: tss1@uakron.edu

2. *Department of Geology*
The University of Akron
Akron, Ohio 44325, USA
E-Mail: tquick@uakron.edu

3: **Deformation Control Technology, Inc.**
7261 Engle Rd., Suite 105
Cleveland, OH 44130
E-Mail: andy.freborg@deformationcontrol.com

Abstract

In this technical paper the results of a recent study aimed at understanding the high cycle fatigue properties and fracture behavior of an alloy steel, a viable candidate for use in performance-critical applications, is presented and briefly discussed. The alloy steel investigated was 9310. The material was evaluated in the as-forged (wrought) and normalized condition. Test specimens of this alloy steel were precision machined and conformed to specifications delineated in the ASTM E8. The as-machined and subsequently polished test samples were cyclically deformed over a range of maximum stress, in the room temperature (T = 25 C), laboratory air environment (Relative Humidity 55 pct), at the load ratios of 0.1 and -1.0. The number of cycles-to-failure was recorded. The specific significance of load ratio on cyclic fatigue life of alloy steel 9310 is presented and differences discussed based on a synergism of the nature of loading, intrinsic microstructural effects, and macroscopic fracture behavior. The fatigue fracture surfaces were examined in a scanning electron microscope to determine the macroscopic fracture mode and to concurrently characterize the intrinsic features on the fatigue fracture surfaces and thus establish the microscopic mechanisms governing failure. The conjoint influence of microstructure, maximum stress and load ratio on cyclic fatigue life and fracture behavior is highlighted.

Key words: Alloy steel, microstructure, cyclic fatigue, load ratio, fatigue life.

Introduction

The economic and social benefits arising from the selection and use of high strength steels for a spectrum of performance-critical and even non-performance critical applications has in recent years gained increased attention. During the same time period, several problems have been identified from the preliminary research and development efforts on high strength alloy steels relating specifically to complex microstructures containing the micro-constituent martensite. These problems include low to inferior fracture toughness, limited ductility, and limited fatigue resistance coupled with inferior resistance to fracture. This provided the much needed interest, incentive and impetus for refining the initial microstructure of the candidate steel as a viable and affordable mechanism for enhancing strength while concurrently ensuring adequate ductility and improved fracture toughness. Recent efforts have focused on using processing techniques as a viable alternative to refine the grain size of the starting metal [1, 2]. This led to the development, emergence and prudent use of mechanical deformation in synergism with heat treatment, or thermo-mechanical processing (TMP), as an effective, efficient and economically affordable alternative to the use of conventional processing in large scale mills or techniques of severe plastic deformation, which are essentially confined to laboratory scale sample dimensions.

A variety of engineering components and structures currently in active use often encounter either variable or constant amplitude fatigue loading, and at times a combination of the two. At any given time, loading on the component or structure of interest can involve high tensile loads. Many of the components often experience a high tensile mean stress [$\sigma_{mean} = \sigma_{maximum} + \sigma_{minimum}/2$] and a high load ratio [$R = \sigma_{minimum} / \sigma_{maximum}$], such as, the pre-tightened bolts, backhoe buckets, and aircraft wings and associated structures during flight. Alloy steel AISI 9310 is often selected as the candidate material for these applications due to its high strength, acceptability ductility, and adequate fatigue resistance. Subsequent to quenching and tempering this alloy steel has martensite as the major micro-constituent in the microstructure [3-5]. However, the intrinsic microstructural features tend to differ considerably even if the post tempering hardness is nearly the same. This is due principally to composition and the resultant type, shape and volume fraction of the key micro-constituents in the microstructure of this steel. Furthermore, the endurance limit or fatigue limit of the alloy steels having near similar composition is easily influenced by the microstructural parameters, such as: (i) size and shape of the primary ferrite grains, (ii) size, shape and volume fraction of the micro-constituent martensite, (iii) presence of both coarse and intermediate size second-phase particles along with fine precipitates, and even (d) mechanical properties to include strength, ductility and fracture toughness. In evaluating the fatigue resistance of this high strength alloy steel, an integral analysis of processing, microstructure, response to loading and resultant fracture behavior is essential. This is in addition to the measurement of hardness while concurrently giving due consideration to both intrinsic microstructural features and mechanical properties as a direct consequence of the neat treatment used to process the steel.

Investigating and understanding the high cycle fatigue behavior of steels, spanning both carbon steels and alloy steels, became the subject of interest following the preliminary study and findings by Naito and co-workers [6]. From their detailed study, Naito and co-workers [6] found that carburized and surface hardened steel does not exhibit a conventional fatigue limit at 10^6 cycles but tends to fail even beyond 10^7 cycles at fairly low values of applied stress. During the subsequent years, specifically in the time period spanning the last three decades, i.e., since the early 1980s, few to several studies have made an effort to evaluate, examine, interpret and record the mechanical behavior of high strength alloy steels relative to baseline microstructures [7-20]. A key finding culminating from these independent studies is that most of these alloy steels tend to fail well below the conventional fatigue limit and in the long life regime beyond 10 7 cycles when cyclically deformed at low values of applied stress. The range of steels that have been systematically and meticulously investigated is rather small and limited essentially to:

(i) the AISI-SAE grades 52100, and
(ii) The Japanese Industrial Standards Steel grades: SUP7, SUP12, SCM 440, SUS304, SUS316, SUS405 and SMn443.

The steels chosen for study, evaluation and examination were the prime candidates chosen and used in systems that often experienced different loading ratios and frequencies during actual service or use. The effect of frequency on high cycle fatigue performance, quantified by fatigue life as a function of maximum stress, was found to be minimal. However, the effect and/or influence of load ratio or stress ratio was observed to play an important role in governing performance of the candidate alloy steel and resultant life of the structure [13].

The objective of this paper is to present and discuss the specific influence of load ratio on high cycle fatigue response and final fracture behavior of alloy steel 9310. The outcome of this research exercise was to identify the conjoint influence of intrinsic microstructural effects and load ratio on fatigue life and resultant fracture behavior. The purpose of the study was to identify and establish the specific role of intrinsic microstructural effects in governing "fatigue life" when subjected to both tension-tension loading and fully reversed tension-compression loading. The microscopic mechanisms and the conjoint influence of intrinsic microstructural features and nature of loading in governing the fracture behavior of this alloy steel at both the macroscopic and fine microscopic levels is established.

Material and Test Specimen Preparation

The alloy steel chosen for this experimental study is AISI 9310. This steel is a preferred candidate for use in gears of helicopters and other related airframe components of use in the aerospace industry. This alloy steel was provided to DCT Inc (Cleveland, Ohio, USA) by the US Army as forged plate. The nominal chemical composition of this alloy steel is given in **Table 1**.

Table 1: Nominal chemical composition of VIM/VAR 9310 STEEL

Element	C	Mn	Si	P	S	Cr	Ni	Mo	Cu	Al	Co	V
9310	0.12	0.57	0.27	0.006	<0.001	1.29	3.33	0.11	0.11	0.05	0.011	<0.01
42CrSi	0.43	0.59	2	-	-	1.33	-	-	-	-	-	-
4140	0.37	1.1	0.35	<0.040	<0.040	1.2	-	0.30	-	-	-	-
X2-M	0.15	0.5	0.9	-	-	5	-	1.4	-	-	-	0.5

This alloy steel (i.e., 9310) was produced by a synergism of vacuum induction melting (VIM) and vacuum arc re-melting (VAR) per AMS 6265. The chemical composition was chosen so as to obtain a stable grain size by the prudent use of thermo-mechanical processing (TMP) coupled with micro-alloying. Presence of carbon helps in refining the grain size while concurrently providing stability to the microstructure. However, the content of carbon is kept low in order to facilitate ease in weldability and is thus maintained to be less than 0.15 percent. The element manganese, like carbon, facilitates in refining the grain size while concurrently increasing hardness at the fine microscopic level. The presence of elements chromium (1.29 weight percent), nickel (3.33 weight percent) and molybdenum (0.11 weight percent) assists in the formation, presence and dispersion of carbide particles, of varying size and shape, in the microstructure. Presence of threes second phase carbide particles contributes to enhancing hardness at the fine microscopic level and strength at the macroscopic level while concurrently resisting softening during heat treatment. The presence of nickel in this alloy steel also aids in lowering the ductile-to-brittle transition temperature while concurrently making it possible to enhance toughness. In fact, addition of nickel facilitates in not only enhancing hardness at the fine microscopic level but also aids in improving the impact strength.

The secondary processing that was used on the as-forged alloy steel is as follows:

(a) This alloy steel was initially vacuum carburized to a depth of 0.038 inch/0.9042 inch.

(b) This was followed by quenching in oil at 140 C to result in a predominantly martensitic microstructure. The presence of pockets of ferrite in the microstructure was made possible by deep freezing this alloy steel in liquid nitrogen.

(c) Finally the alloy steel was double tempered at 350 F.

Cylindrical test specimens conforming to specification outlined in ASEM E-8 (Tension testing of metallic materials) [21] and E-466-10 (Stress amplitude controlled high cycle fatigue test on Metallic Materials) [22], were precision machined from the as-provided forged plate of alloy steel 9310. The stress axis of the machined test specimens was parallel to the longitudinal direction of the as-provided plate stock. At the gage section, the test specimens measured 12.5 mm (0.5 inch) in length and 3.125 mm (0.125 inch) in diameter. To minimize the effects and/or contributions from both surface irregularities and finish, final surface preparation was achieved by mechanically polishing the gage section of all test specimens (tensile plus cyclic fatigue) using progressively finer grades of silicon carbide impregnated emery payer to remove any and all of the circumferential scratches and surface machining marks.

Characterization of Initial Microstructure

Samples were taken from the mid-section of the as-provided alloy steel 9310 stock. Initial characterization of the as-received microstructure was done using a low magnification optical microscope. The samples taken from both the longitudinal and transverse directions of the as-provided alloy steel stock were mounted in bakelite. The mounted samples were initially wet ground on progressively finer grades of silicon carbide impregnated emery paper using copious amounts of water both as a lubricant and as a coolant. Subsequently, the ground sample was mechanically polished using 5-micron alumina (Al_2O_3) powder and 1-micon alumina (Al_2O_3) powder suspended in distilled water as the lubricant. The polished samples were etched using nital reagent, i. e, a solution mixture of 5-ml nitric acid and 95 ml of ethanol. The polished and etched surfaces if the samples were observed in an optical microscope and photographed using standard bright field illumination technique.

Mechanical Testing

The tensile and cyclic fatigue tests were performed on a fully automated, closed-loop servo-hydraulic mechanical test machine (INSTRON: Model 8500 plus) equipped with a 100 KN load cell. The stress amplitude controlled high cycle fatigue tests were performed in the room temperature (25 C) laboratory air environment (Relative Humidity of 55 pct) using a sinusoidal waveform and at stress ratios (R = minimum stress / maximum stress) of 0.1 and -1.0. The stress amplitude controlled fatigue tests were conducted at a constant frequency of 5 Hz. At the two chosen load ratios, representative of tension-tension loading and tension-compression loading, the fatigue tests were conducted over a range of stress amplitudes to establish the variation of maximum stress ($\sigma_{maximum}$) with fatigue life (N_f). The test data obtained from the stress-amplitude controlled tests was used to obtain the following relationships:

(i) Variation of maximum stress ($\sigma_{maximum}$) with fatigue life (N_f)
(ii) Variation of maximum elastic strain ($\sigma_{maximum}$ /E) with fatigue life (N_f)
(iii) Variation of the ratio of maximum stress /yield stress [$\sigma_{maximum}$ / $\sigma_{yield\ stress}$] with fatigue life (N_f)
(iv) Variation of ratio of maximum stress /ultimate tensile strength [$\sigma_{maximum}$ / σ_{UTS}] with fatigue life (N_f).

The graphs depicting the above relationships are shown in **Figure 4** to **Figure 7** for the longitudinal test specimens that were cyclically deformed at the two load ratios representative of tension-tension loading and tension-compression loading, The influence of load ratio (or stress ratio) on fatigue response of this alloy steel (9310) can be inferred from these figures.

Analysis of Damage and Failure

Fracture surfaces of the cyclically deformed and failed test specimens were comprehensively examined in a scanning electron microscope over a range of allowable magnifications to help establish the following:
(a) The macroscopic fracture mode.
(b) The fine scale topography and features on the fatigue fracture surface for the purpose of establishing the microscopic mechanisms contributing to failure by fracture.
Samples for observation in the scanning electron microscope (SEM) were obtained from the deformed and failed specimens by sectioning parallel to the fracture surface.

Results and Discussion

Initial Microstructure.

Optical microstructure of the as-provided forged plate of alloy steel 9310 in the three orientations, longitudinal, long transverse and short transverse, is shown in **Figure 1**. The observed micrographs in the three orientations are quite typical of low-carbon high strength alloy steel having a combination of fine acicular martensite and retained austenite. The presence of sizeable quantity of alloying elements in this steel resulted in noticeable volume fraction of the micro-constituent martensite. The martensite observed in the microstructure had a needle-like or acicular morphology and is thus a preferred site for high "local" stress concentration. In addition to the two primary phases, or micro-constituents: martensite and ferrite, dispersed through the microstructure were an observable population of fine carbide particles. The particles result as a direct consequence of the high alloy content in this steel. The presence and morphology of the micro-constituent martensite is governed by both composition of the steel and the secondary processing technique used to engineer this steel. These intrinsic features, i.e., ferrite and martensite, in combination with the presence and dispersion of the fine carbide particles exert a positive influence on hardness, tensile strength, response when subjected to cyclic loading and resultant fracture behavior at both the macroscopic level and fine microscopic level. A triplanar optical micrograph showing the microstructure in the three orthogonal orientations is shown in **Figure 2.**

Figure 1. **Optical micrographs showing the key microstructural constituents on the polished and etched surface of alloy steel 9310:**
(a) Longitudinal orientation, (b) Long-Transverse orientation, and
(c) Short Transverse orientation.

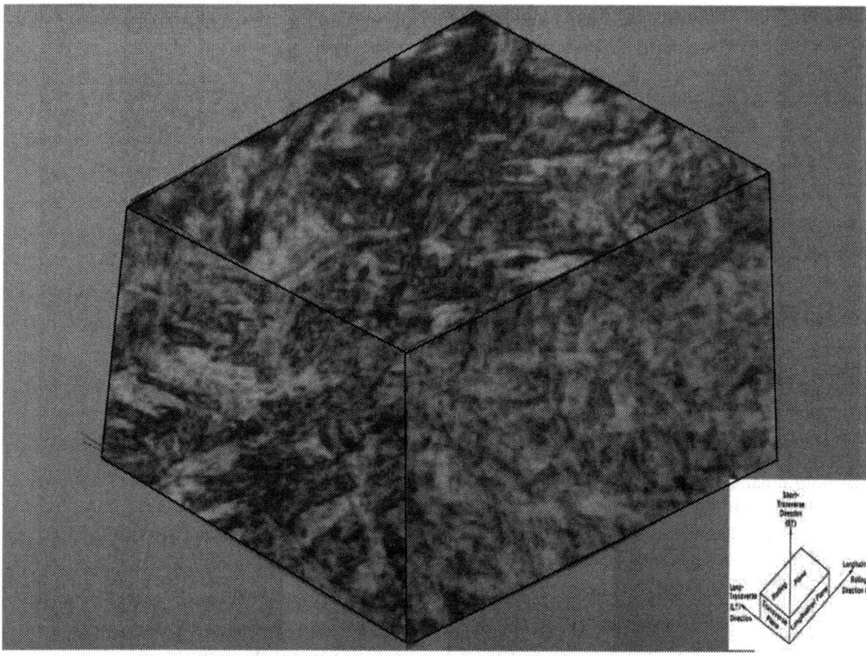

Figure 2: **Triplanar optical micrograph illustrating the microstructure of alloy steel 9310 forged plate in the three orthogonal directions.**

Tensile Properties

The room temperature tensile properties of the as-provided alloy steel 9310 are summarized in **Table 2**. The results obtained from duplicate tests and the average value is reported. The yield strength of this alloy steel was 1040 MPa in the longitudinal direction. The ductility quantified by elongation over the 12.5 mm gage length was 14 percent. The reduction in test specimen cross-section area, a measure of ductility, was 59 percent. The ultimate tensile strength of this alloy steel is 1239 MPa, which is 19.0 percent higher than the yield strength providing clear and convincing indication of the occurrence of strain hardening beyond yield. The engineering stress versus engineering strain curve of this steel is shown in **Figure 3**.

Table 2: **A Compilation Of The Room Temperature (T = 25°C) Tensile Properties Of Alloy Steel 9310 Taken In The Longitudinal Orientation.**

Orientation	Elastic Modulus		Yield Strength		UTS		Elongation GL=1" (%)	Reduction In Area (%)
	ksi	GPa	ksi	MPa	ksi	MPa		
9310	33967.9	234.2	150.7	1039.6	179.7	1239	13.9	59
42CrSi	33503.7	231	261.5	1803	296.9	2047	15.8	2.9
4140	33068.6	228	96.5	665	123.3	850	13	25
X2-M	34228.9	236	165.1	1138	175.1	1138	16	55

For both tensile tests this alloy steel exhibited continuous yielding behavior. The occurrence of continuous yielding is ascribed to be due to the presence of unpinned dislocations introduced in the soft ferrite regions in the microstructure during highly 'localized' plastic deformation. The unpinned dislocations are located primarily at the interface of the soft ferrite and the hard martensite and are assumed to be mobile during the early stages of plastic deformation. The amount and/or volume fraction of the harder micro-constituent, i.e., martensite, present in the microstructure of this alloy steel does exercise a control over the density of dislocations. The velocity of the mobile dislocations and their resultant movement through the microstructure is controlled by not only the amount of the soft ferrite phase but also by its location and distribution, or in essence the volume fraction of ferrite in the final microstructure.

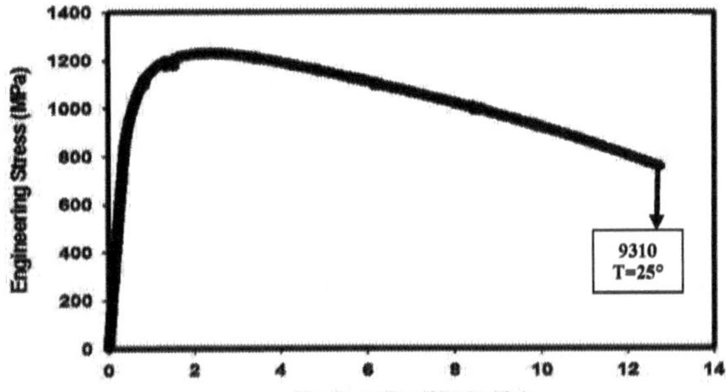

Figure 3: **Variation of engineering stress (MPa) with engineering strain (percent) for alloy steel 9310.**

Cyclic Stress Amplitude Controlled High Cycle Fatigue Response

At ambient test temperature (25°C) the results of the stress amplitude controlled fatigue tests are shown in **Figure 4** as the variation of maximum stress ($\sigma_{maximum}$) with fatigue life (N_f). The maximum stress with fatigue life curve shows a trend of increase in fatigue life with a gradual decreased in the value of maximum stress, a trend shown by most metals and even their composite counterparts. The maximum stress versus fatigue life (N_f) response of this AISI 9310 alloy steel when deformed under the influence of tension-tension loading (i.e. R = 0.1) is noticeably better than the response observed when deformed under fully-reversed loading, i.e. R = -1.0, which is a trend expected and observed in most metals and their composite counterparts. The endurance limit of this alloy steel when cyclically deformed at load ratio of 0.1 is 800 MPa and over 200 percent better than the endurance limit when deformed at load ratio of -1.0, which is 400 MPa. A load ratio of -1.0, representative of fully-reversed loading, is the toughest or most severe loading condition that a metal can be subjected to during cyclic loading and an endurance limit of 400MPa shown by this alloy steel reveals noticeably good cyclic fatigue resistance,

To facilitate a better understanding of the response of this 9310 alloy steel from the standpoint of design engineering, the variation of maximum stress ($\sigma_{maximum}$) with fatigue life (N_f) is re-plotted in terms of the variation of maximum elastic strain [$\sigma_{maximum}$ / E] to take into account the elastic modulus if the test specimen, as a function of fatigue life. This is shown in **Figure 5** and reveals a linear trend. A noticeable difference in behavior between the two load ratios, i.e., R = 0.1 and R = -1.0, is observed at all values of maximum elastic strain. At a given value of the maximum

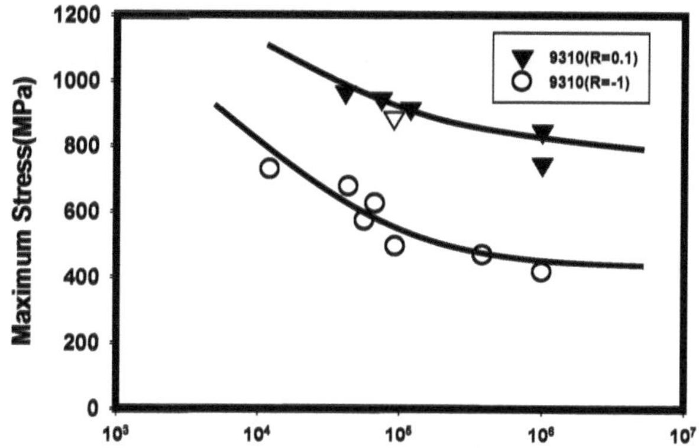

Figure 4. **Influence of load ratio on variation of maximum stress (σmaximum) with fatigue life (N_F) for alloy steel 9310 cyclically deformed at room temperature (25°C).**

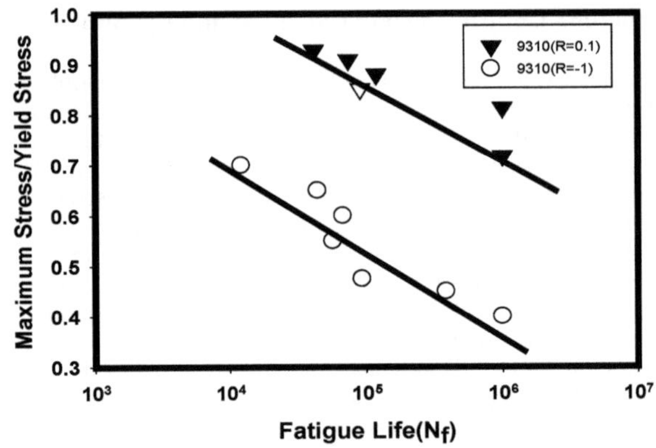

Figure 5. **Influence of load ratio on variation of maximum elastic strain ($\sigma_{maximum}$ /E) with fatigue life (N_f) for alloy steel 9310 cyclically deformed at room temperature 25°C.**

elastic strain (σ $_{maximum}$ /E), the observed degradation in cyclic fatigue life (N_f) of the testy specimen deformed under fully reversed tension-compression loading (R = -1) is noticeably large and distinctly noticeably when compared to the test specimen that was deformed under tension-tension loading (i.e. R =0.1). The degradation is as high as 300 pct at all values of the ratio of maximum elastic strain. Also, for the longitudinal orientation the variation of ratio of maximum

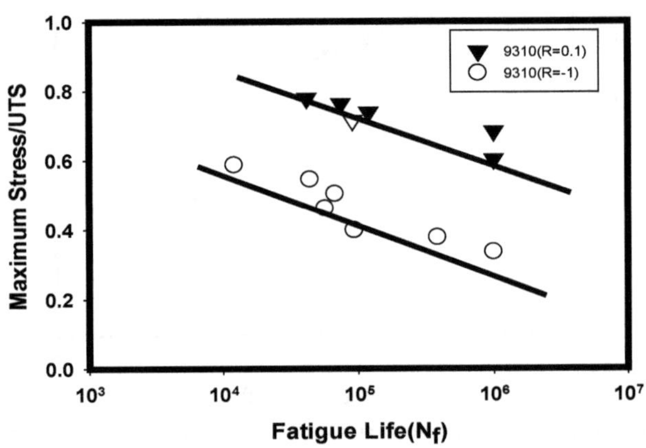

Figure 6. **The variation of ratio of maximum stress to yield stress [$\sigma_{maximum}$ / σ_{YS}] with fatigue life (N_F) for alloy steel 9310 cyclically deformed at room temperature (25°C).**

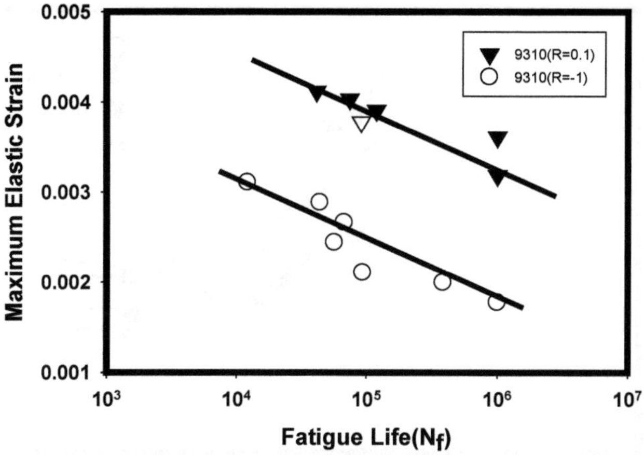

Figure 7: The variation of the ratio of maximum stress to ultimate tensile strength [$\sigma_{maximum}$ / σ_{UTS}] with fatigue life (N_f) for alloy steel 9310 cyclically deformed at room temperature (25°C).

stress/yield stress as a function of fatigue life (N_f) revealed the fatigue response under fully reversed loading (R = -1) to be noticeably inferior than the tension-tension loading (R = 0.1) (**Figure 6**.). A similar trend is noticed in examining the variation of the ratio of maximum stress to ultimate tensile strength [$\sigma_{maximum}$/UTS] with fatigue life as shown in **Figure 7**. The observed degradation in cyclic fatigue life and resultant fatigue response is due to the conjoint and mutually interactive influences of nature of loading, intrinsic microstructural effects, matrix deformation characteristics and macroscopic mode of fracture assisted with the microscopic mechanisms controlling fracture at the fine microscopic level.

Comparing the maximum stress versus fatigue response of this alloy steel 9310 with other candidate alloy steels, cyclically deformed under conditions of fully reversed loading, i.e., R = -1, is as shown in **Figure 8.** Over the entire range of maximum stress this steel reveals inferior fatigue life when compared with the emerging steel X2M an experimental steel 42SiCr. However, little to minimal difference in stress versus fatigue life response was evident with the widely used commercial alloy steel 4140. The noticeably improved maximum stress-fatigue life response of the experimental alloy steel 42CrSi and emerging steel X2M can be ascribed to their higher yield strength than the candidate alloy steel 9310 and the commercial and widely chosen and used steel 4140.

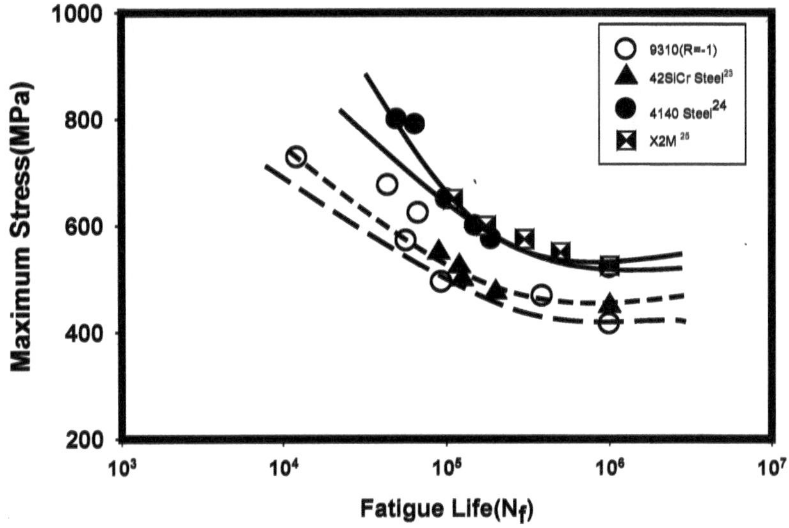

Figure 8: **A comparison of the maximum stress versus fatigue life response of alloy steel 9310 with other candidate alloy steels, at constant load ratio of -1.**

Cyclic Fracture Behavior

A comprehensive examination of the fracture surfaces of the cyclically deformed and failed test specimens was done using a FEI Quanta 200 scanning electron microscope (SEM). Low magnification observation was used to identify:

(a) Regions specific to the onset of crack initiation and damage during cyclic loading and final fracture behavior (region of overload), and

(b) At gradually higher allowable magnifications of the SEM to critically observe the regions of both fatigue and overload so as to identify the following:

  (1) Location of microscopic crack initiation,

  (2) Nature and extent of early microscopic crack growth through the microstructure of this alloy steel, and

  (3) Other fine scale features on the fracture surface.

Representative fracture features of the test specimen cyclically deformed at the two load ratios and deformed at (i) high maximum stress and resultant short fatigue life, and (ii) low maximum stress and resultant enhanced fatigue life are shown in **Figure 9-12.**

Load Ratio of -1.0

Scanning electron micrographs of the fatigue fracture surface of the failed 9310 alloy steel test sample that was cyclically deformed at a maximum stress of 727.7 MPa with a resultant fatigue life of 12,191 cycles revealed global or macroscopic failure to be essentially normal to the far-field stress axis and comprising of two distinct regions clearly representative of fatigue and overloads (**Figure 9-a**). The region of early microscopic crack growth upon observation at higher allowable magnifications of the scanning microscope revealed the presence and distribution of fine microscopic voids of varying size and shape (**Figure 9-b**). Careful high magnification observation of this region and gradually approaching the domain of stable crack growth revealed pockets of shallow striations reminiscent of micro-plastic deformation intermingled with fine microscopic cracks parallel to the major stress axis and indicative of locally occurring brittle failure (**Figure 9-c**). At higher magnifications the region immediately prior to overload revealed (i) an observable population of microscopic voids intermingled with dimples of varying size, and (ii) pockets of well-defined striations features that are indicative of locally occurring ductile failure mechanisms (**Figure 9-d**).

The test sample that was cyclically deformed at a lower maximum stress of 467.8 MPa with a resultant fatigue life of 385,763 cycles is as shown in **Figure 10**. Overall morphology of failure was flat and essentially normal to the far-field stress axis (**Figure 10-a).** High magnification observation of the fracture surface revealed the transgranular regions to be inlaid with shallow striations indicative of the occurrence of micro-plastic deformation (**Figure 10-b**). At higher magnification was evident the profile and nature of macroscopic cracks intermingled with very fine microscopic cracks **(Figure 10-c**). In the region of stable crack growth immediately prior to unstable crack growth and overload an observable population of striations was seen providing evidence of micro-plastic deformation and stable crack growth through much of the alloy microstructure for this specimen exhibiting long cyclic fatigue life **(Figure 10-d**).

Load Ratio of 0.1

Scanning electron micrographs of the test sample that was cyclically deformed at a maximum stress of 964.2 MPa with a resultant fatigue life of 41,584 cycles is shown in **Figure 11**. Overall morphology of failure revealed distinct regions of fatigue and overload. The region of early crack growth revealed both macroscopic cracks intermingled with fine microscopic cracks (**Figure 11-b**). Essentially the region of early microscopic crack growth was flat and inlaid with an observable population of cracks to include few macroscopic and many fine microscopic. All of the cracks were oriented in the direction of the major stress axis. In the region of unstable crack growth and overload was evident shallow dimples of varying size intermingled with fine microscopic cracks (**Figure 11c**). These features are representative of the locally occurring ductile and brittle failure mechanisms. In the region of overload when observed at high magnification revealed a noticeable population of microscopic voids and their coalescence to form a fine microscopic crack **(Figure 11-d**), the key features indicative of both ductile and brittle failure mechanisms occurring at the fine microscopic level.

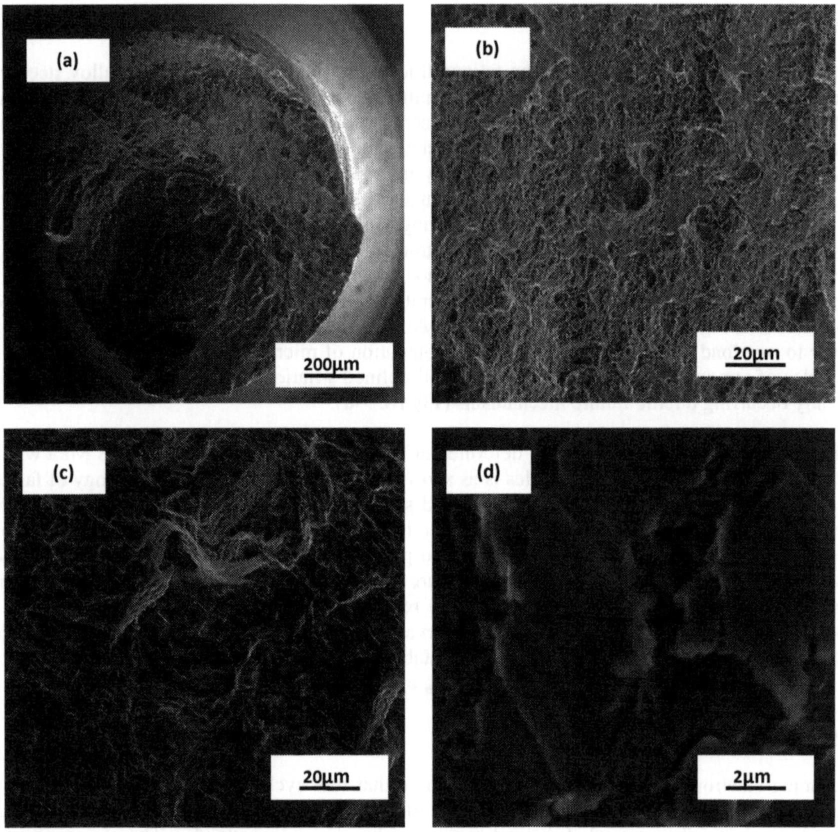

Figure 9: **Scanning electron micrographs of the fatigue fracture surface of the 9310 sample cyclically deformed at load ratio of -1 at a maximum stress of 727.7 MPa, resultant fatigue life of 12,191 cycles, showing:**

(a) Overall morphology of failure
(b) High magnification observation of the region of early microscopic crack growth shoeing isolated pockets of fine microscopic voids.
(c) The presence of pockets of shallow striations intermingled with very microscopic cracks in the region of stable crack growth.
(d) High magnification observation of the fracture surface immediately prior to unstable crack growth showing morphology and clarity of striations.

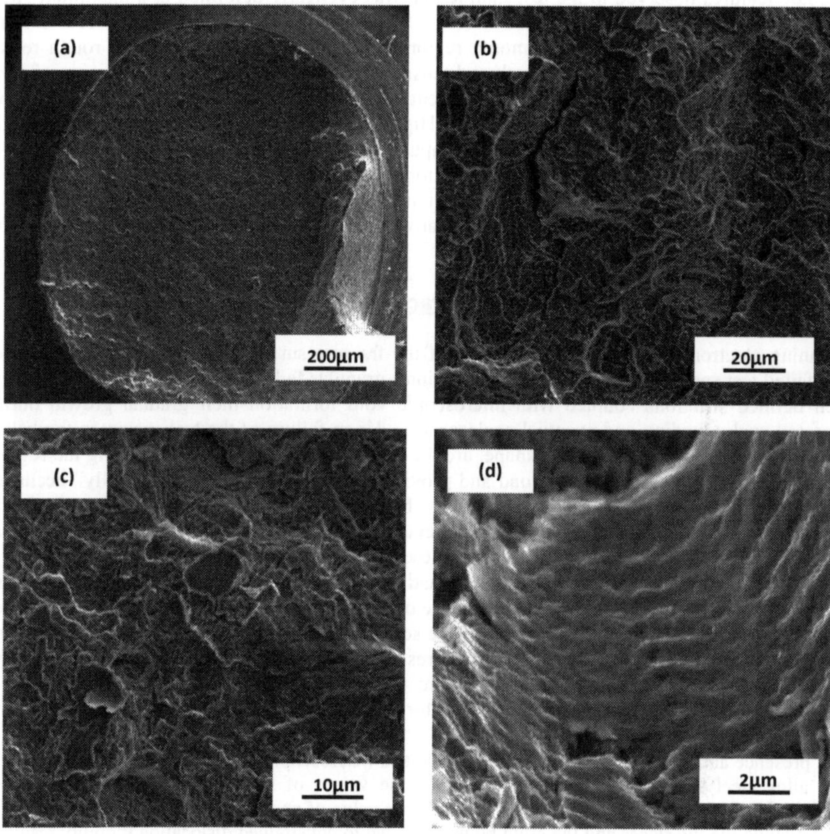

Figure 10. **Scanning electron micrographs of the fatigue fracture surface of the 9310 sample cyclically deformed at load ratio of -1 at a maximum stress of 467.8 MPa, resultant fatigue life of 385,763 cycles, showing:**

(a) Overall morphology of failure
(b) The transgranular fracture region showing pockets of shallow striations dispersed through the surface
(c) Array of microscopic cracks and macroscopic cracks parallel to the major stress axis
(d) Well defined, yet shallow striations, in the region of stable crack growth

The key features observed in the test sample that was cyclically deformed at a maximum stress of 884.7 MPa and a resultant fatigue life of 91,120 cycles is shown in **Figure 12**. Overall morphology of failure revealed a smooth region representative of fatigue and a rough region representative of unstable crack growth and tensile overload (**Figure 12**a). The region of early microscopic and stable crack growth was essentially smooth but inlaid with both macroscopic and fine microscopic cracks **(Figure 12b)**. High magnification observation of the region of stable crack growth revealed an observable population of striations, which are representative of the occurrence of "localized" micro-plastic deformation **(Figure 12 c)**. The region of overload when observed at higher allowable magnifications of the SEM revealed a noticeable population of dimples intermingled with fine voids, indicative of predominantly ductile failure mechanisms at the fine microscopic level (**Figure 12d**).

Micro-mechanisms Governing Cyclic Fracture Behavior.

Scanning electron microscopy observations of the fracture surfaces of the cyclically deformed and failed test specimen's revealed pockets of inter-granular failure intermingled with pockets of well defined striations coupled with microscopic void formation their gradual growth during repeated cyclic loading and eventual coalescence. Upon failure of the test specimen the halves of these voids, of varying size and shape, are the shallow dimples observed covering the regions of unstable crack growth and overload and provides a representative of the "locally" occurring failure mechanisms. The presence of cracks, both fine microscopic and macroscopic, in the region of early microscopic crack growth and even the region on stable crack growth is assumed to be due to the relative ease of nucleating these fine microscopic cracks by the occurrence of decohesion at the interfaces of the soft and plastically deforming ferrite matrix and the hard and essentially elastically deforming martensite (the dominant phase). The occurrence of decohesion at both the region of the interface between the soft and hard micro-constituent, and at the grain boundary regions is responsible for the features indicative of locally brittle failure at the fine microscopic level while matrix microstructure (traces of ferrite) and its intrinsic softness is responsible for the locally occurring ductile failure mechanisms evident by voids and dimples.

The presence and severity of the fine microscopic and macroscopic cracks on the fracture surface of this alloy steel specimen cyclically deformed at the values of maximum stress is a function of magnitude of maximum applied stress. At the lower value of applied maximum stress the cracking at and along the interfaces and the grain boundaries due to constrained dislocation mobility resulting in dislocation pile up and concomitant local stress concentration did not initiate immediately upon cycling. A substantial time period of cyclic loading was required that favored well defined and distinct region of stable crack growth, which was covered with fine striation like features indicative of micro-plastic deformation prior to the inset of micro-cracking and their gradual growth and eventual coalescence to result in macroscopic cracks

At the higher value of maximum stress the cracking initiated early during cyclic deformation and both fine microscopic and macroscopic cracks were evident through the fatigue fracture surface. Also, the initiation of fine microscopic cracks is facilitated by the initiation, gradual growth and eventual coalescence of the microscopic voids. Following the initiation of the micro-voids at grain boundaries an. second phase particles dispersed through the microstructure their progressive growth results as a direct consequence of micro-plastic deformation occurring at the

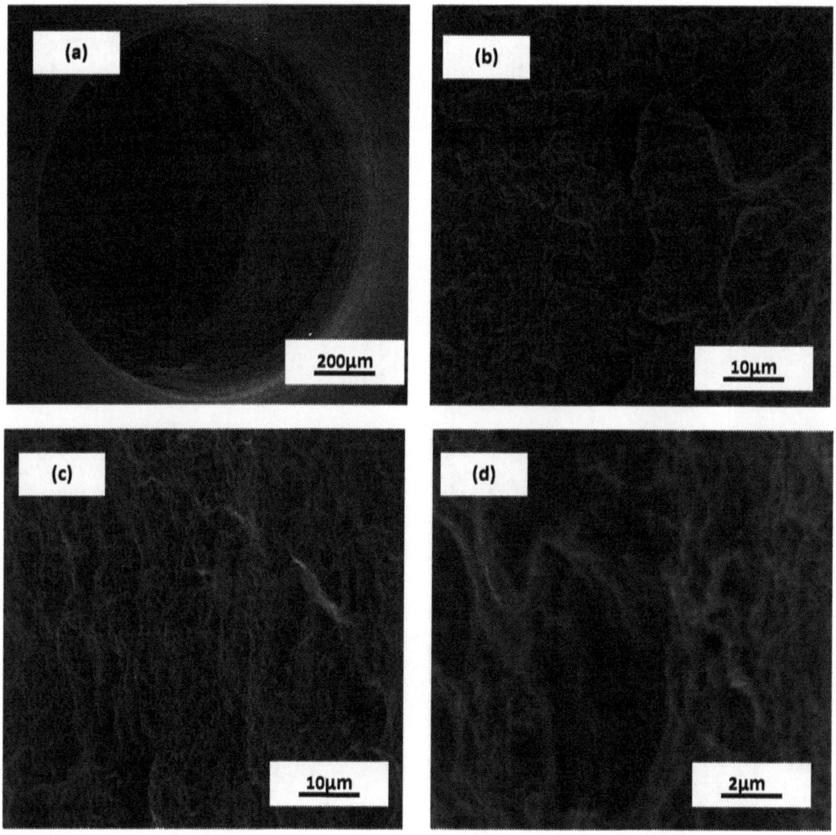

Figure 11. **Scanning electron micrographs of the fatigue fracture surface of the 9310 sample cyclically deformed at load ratio of 0.1 at a maximum stress of 964.2 MPa, resultant fatigue life of 41,584 cycles, showing:**

- **(a) Overall morphology of failure.**
- **(b) The region of fracture surface in the domain of early crack growth**
- **(c) High magnification observation of (b) showing key and intrinsic features covering the fracture surface.**
- **(d) The region of overload failure showing a population of fine microscopic voids, void coalescence to form microscopic cracks and intermingled with pockets of dimples**

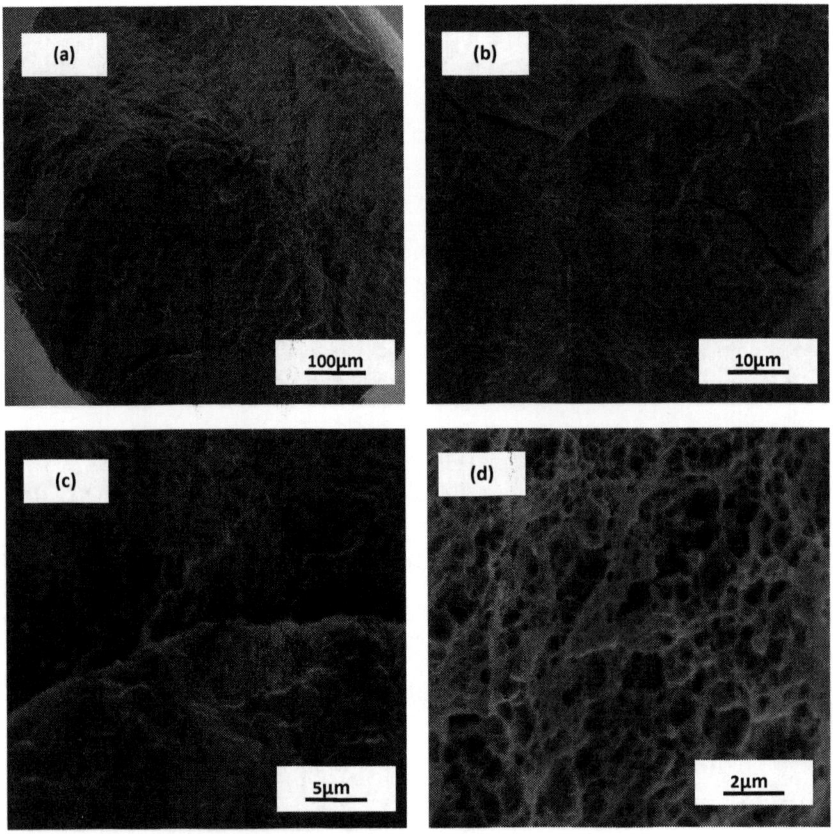

Figure 12. **Scanning electron micrographs of the fatigue fracture surface of the 9310 sample cyclically deformed at load ratio of 0.1 at a maximum stress of 884.7 MPa, resultant fatigue life of 91,120 cycles, showing:**

(a) Overall morphology of failure of the test specimen.

(b) High magnification observation of the region of initiation and early microscopic crack growth.

(c) The region of fracture surface transition between stable and unstable crack growth

(d) Healthy array of dimples and fine voids covering the overload fracture surface.

local level. The gradual accumulation of micro-plastic damage is exacerbated by stress-microstructure interactions that occur during each cycle of loading. It is this irreversible plastic deformation that is responsible for the initiation, progressive growth through the alloy steel microstructure and eventual coalescence of the fine microscopic cracks with each other and other propagating macroscopic cracks.

The presence of fine voids of varying size and shape does transforms the polycrystalline alloy steel to a composite at the microscopic level. Subsequent to their initiation at the coarse second phase particles dispersed through the microstructure, their incremental growth during repeated cyclic loading culminating in their eventual coalescence with each other results in the initiation of a fine microscopic crack while concurrently facilitating extension of the macro crack through the alloy steel micro-structure. At the second phase particles plastic deformation is localized due to the pile-up of dislocations during repeated cyclic loading thus facilitating conditions for the initiation of fine microscopic voids.

Conclusions

The results of a study aimed at understanding the specific influence of nature of loading, as quantified by load ratio, on high cycle fatigue response and resultant fracture behavior of alloy steel 9310 in the longitudinal orientation provides the following key findings:

1. Light optical microscopy of this high strength alloy steel revealed a sizeable volume fraction of martensite intermingled with pockets of ferrite. The alloying elements present in this steel are responsible for the high volume fraction of martensite. Presence of the micro-constituent martensite was needle-like or acicular.

2. The yield strength of this alloy steel in the longitudinal orientation was high and 1040 MPa. The ultimate tensile strength was 19 percent higher than the yield strength, indicating the occurrence of strain hardening beyond yield. For this high strength this alloy steel exhibited 14 percent elongation and 59 pct reduction in cross-section area.

3. The variation of maximum stress with fatigue life followed the trend expected for most metals and their composite counterparts. At equivalent values of maximum stress the fatigue life of the specimen deformed at load ratio of -1 was noticeably inferior to life of the test specimen deformed under tension-tension loading at load ratio of 0.1.

4. Over the entire range of maximum stress the 9310 alloy steel revealed inferior fatigue life when compared with the emerging steel X2M an experimental steel 42SiCr, and minimal difference in stress versus fatigue life response with the widely used commercial alloy steel 4140. The noticeably improved maximum stress-fatigue life response of the experimental alloy steel 42CrSi and emerging steel X2M can be ascribed to their higher yield strength.

5. Scanning electron microscopy observation of the fracture surface of the deformed and failed specimens revealed features representative of the level of maximum stress and resultant fatigue life. The region of fatigue comprised of a very short initiation site, early microscopic crack growth and a short, yet distinct domain of stable crack growth through the microstructure of alloy steel. Both macroscopic and fine microscopic cracks were evident in the region of early microscopic crack growth. Pockets of well defined striations were evident in the region of stable crack growth. The overload region revealed voids and dimples reminiscent of locally occurring ductile failure mechanisms.

References

1. Howe AA, et al. ECSC contract 7210.PR/167. Luxembourg: EC, 2000.
2. Song R, Ponge D, Raabe D, Speer JG, Matlock DK,.." ***Materials Science and Engineering A***", Vol. 441, 2006, pp. 1-11.
3. DIN handbook 404, Iron and steel, quality standards' 4. 3rd ed. Germany: Deutsches Institut Fur Normung E.V., 2002.
4. ASM Metals Handbook. Heat Treating, Metals park, vol. 4. Ohio: American Society for Metals, 1991.
5. Chandler H. *Heat treating guides*. Metals Park, Ohio: American Society for Metals; (1996).
6. Naito T, Ueda H, Kikuchi M." Fatigue behaviour of carburized steel with internal oxides and non-martensitic microstructure near the surface". ***Metallurgical Transactions A*** (1984),15 A, pp 1431–1436.
7. Marines I, Dominguez G, Baudry G, Vittori J-F, Rathery S, DoucetmJ-P, et al. Ultrasonic fatigue tests on bearing steel AISI-SAE 52100 at frequency of 20 and 30 kHz. ***International Journal of Fatigue***, 2003, 25, pp. 1037–46.
8. Wang QY, Bathias C, Kawagoishi N, Chen Q." Effect of inclusion on subsurface crack initiation and gigacycle fatigue strength", ***International Journal of Fatigue***, (2002), 24, pp. 1269–1274.
9. Furuya Y, Matsuoka S, Abe T. "Inclusion-controlled fatigue properties of 1800 MPa-class spring steels. " ***Metallurgical and Materials Transaction A***, 2004, 35A, 3737–3744.
10. Abe T, Furuya Y, Matsuoka S." Gigacycle fatigue properties of 1800 MPa class spring steel, " ***Fatigue Fracture of Engineering Materials and Structures***, 2004, 27, pp. 159-167.
11. Furuya Y, Matsuoka S. "Gigacycle fatigue properties of a modified ausformed Si–Mn steel and effects of microstructure". ***Metallurgy and Materials Transactions*** *A* (2004), 35A, pp. 1715–1723.
12. Furuya Y, Abe T, Matsuoka (S. 10^{10}-cycle fatigue properties of 1800 MPa-class JIS-SUP7 spring steel,) "***Fatigue and Fracture of Engineering Materials and Structures***, 2003, 26, pp. 641–645.
13. Furuya Y, Matsuoka S, Abe T, Yamaguchi K. "Gigacycle fatigue properties for high-strength low-alloy steel at 100 Hz, 600 Hz, and 20 kHz". ***Scripta Materialia***, 2002, 46, pp.157–162.
14. Itoga H, Ko H-N, Tokaji K, Nakajima M. Effect of inclusion size on step-wise S–N characteristics in high strength steels. In: VHCF-3: *Proceedings of the third international conference on very high cycle fatigue;* 2004, pp. 633–640.
15. Tokaji K, Ko H-N, Nakajima M, Itoga H. Effects of humidity on crack initiation mechanism and associated S–N characteristics in very high strength steels, ***Materials and Science Engineering A***, 2003, A345:197–206.
16. Melander A, Larsson M. The effect of stress amplitude on the cause of fatigue crack initiation in a spring steel. ***International Journal of Fatigue***, 1993, 15(2): 119–31.
17. Larsson M, Melander A, Nordgren "A. Effect of inclusions on fatigue behaviour of hardened spring steel". ***Materials Science and Technology***, 1993, 9, pp. 235–45.

18. Murakami Y, Yokoyama NN, Nagata J. Mechanism of fatigue failure in ultralong life Regime", ***Fatigue and Fracture of Engineering Materials and Structures***, 2002, 25, pp.735–46.
19. Murakami Y, Nomoto T, Ueda T, Murakami Y. On the mechanism of fatigue failure in the superlong life regime (N > 10^7 cycles). Part I: influence of hydrogen trapped by inclusions, "***Fatigue and Fracture of Engineering Materials and Structures***, 2000, 23, pp. 893–902.
20. Murakami Y, Nomoto T, Ueda T, Murakami Y. "On the mechanism of fatigue failure in the superlong life regime (N > 10^7 cycles). Part II: a fractographic investigation". ***Fatigue and Fracture of Engineering Materials and Structures***, 2000, 23, pp. 903–910.
21. American Society for Testing and Materials: Standard E-8-10: "*Conducting Tension Test on Metallic Materials*," American Society for Testing and Materials (ASTM), Philadelphia, PA, USA, 2010.
22. American Society for Testing and Materials: Standard E-466-02: "*Conducting Constant Amplitude Axial Fatigue Tests on Metallic Materials*," American Society for Testing and Materials (ASTM), Philadelphia, PA, USA, 2002.
23. Ivo Černýa, Dagmar Mikulováa et al: "Fatigue properties of a low alloy 42SiCr steel heat treated by quenching and partitioning process", ***Procedia Engineering*** 10, 2011, pp. 3310–3315.
24. K. Padillaa, A. Vela´squeza, et al."Fatigue behavior of a 4140 steel coated with a NiMoAl deposit applied by HVOF thermal spray", ***Surface and Coatings Technology*** 150, 2002, pp. 151–162
25. D.P Davis and Jenkins, "Influence of stress and environment on the fatigue strength and failure characteristics of case carburized low alloy steels for aerospace applications". ***International Journal of Fatigue***, 2012, (In press).

Acknowledgements:

The authors (TSS, MK and AMF) extend sincere thanks, gratitude and appreciation to The **U.S. Navy** for partial support of this research to **Deformation Control Technology, Inc** (Cleveland, Ohio, USA).

Fatigue of Materials II Advances and Emergences in Understanding
Edited by: T.S. Srivatsan, M. Ashraf Imam, and R. Srinivasan
TMS (The Minerals, Metals & Materials Society), 2013

ULTRASONIC CORROSION FATIGUE BEHAVIOR OF HIGH STRENGTH AUSTENITIC STAINLESS STEELS

R. Ebara[1], Y.Yamaguchi[2], D.Kanei[3], Y.Yamamoto[4]

1: ***Institute of Materials Science and Technology***
Fukuoka University,
8-19-1, Nanakuma, Jonan-ku
Fukuoka-city, 814-0180, **Japan**
E-Mail:ebara@fukuoka-u.ac.jp

2,3,4: ***Student, Department of Mechanical System Engineering***
Hiroshima Institute of Technology
2-1-1, Miyake, Hiroshima, 731-5193, **Japan**

Keywords: Ultrasonic Corrosion Fatigue, YUS270, SUS304N2, 2.5% Mo Containing High Strength Austenitic Stainless Steel, Corrosion Pit, Striation

Abstract

Ultrasonic corrosion fatigue tests were conducted for high strength austenitic stainless steels such as YUS270 and SUS304N2 in 3%NaCl aqueous solution. The reduction of giga-cycle corrosion fatigue strength of YUS270 and SUS304N2 was not observed at all, while the reduction of corrosion fatigue life was observed at higher stress amplitude. Corrosion pit was observed on corrosion fatigue crack initiation area. Striation was predominantly observed on crack propagation area in air and in 3% NaCl aqueous solution. The reduction of corrosion fatigue strength of high strength austenitic stainless steels such as YUS270 and SUS304N2 is due to the corrosion pit formation at corrosion fatigue crack initiation area. It can be concluded that the higher the ultimate tensile strength of austenitic stainless steels the higher the giga-cycle corrosion fatigue strength in 3%NaCl aqueous solution is.

Introduction

Corrosion fatigue problems of austenitic stainless steels are still very important matter in life prediction and design of various components. However long term corrosion fatigue strength and mechanism of corrosion fatigue crack initiation behavior of austenitic stainless steels are not well understood [1]. Hirakawa and Kitaura reported that corrosion fatigue strength of SUS304 suddenly drops after $2x10^6$ cycles and the reduction rate became 27 percent at around 10^7 cycles. The reason of this sudden reduction of corrosion fatigue strength at long term was attributed to corrosion fatigue crack initiation from corrosion pit[2].This phenomenon suggest that investigation on corrosion fatigue crack initiation is absolutely necessary to evaluate long term corrosion fatigue strength of austenitic stainless steels. One of the authors conducted corrosion fatigue tests of base metal and laser welded specimens for newly developed 2.5%Mo containing high strength austenitic stainless steel in 3% NaCl aqueous solution. It was clarified that reduction of corrosion fatigue strengths of base metal and laser welded specimens for heat treated material and TMCP made material in 3% NaCl aqueous solution were not significant. The important role of corrosion pit in corrosion fatigue crack initiation process was demonstrated by electrochemical noise measurement and surface and fracture surface observation[3].

It is well established that ultrasonic corrosion fatigue testing brought us rapidity and effectiveness in corrosion fatigue testing to obtain corrosion fatigue strength and corrosion fatigue crack propagation rate of structural materials[4]. The ultrasonic corrosion fatigue testing data on 12Cr martensitic stainless steel was effectively used for evaluation of corrosion fatigue behavior of steam turbine blade [5].The authors reported on ultrasonic corrosion fatigue behavior of austenitic stainless steels such as SUS304,SUS316 and 2.5%Mo containing austenitic stainless steels and discussed on corrosion fatigue crack initiation behavior of these austenitic stainless steels is on the basis of corrosion pit observed at crack initiation area[6].

In this paper ultrasonic corrosion fatigue behavior of high strength austenitic stainless steels is evaluated on the basis of the giga-cycle corrosion fatigue data of high strength austenitic stainless steels such as YUS270,SUS304N2 and 2.5% Mo containing austenitic stainless steel[5].

Experimental procedure

The tested materials were YUS270 (Super stainless steel) and SUS304N2 (High Nitrogen Austenitic stainless steel). Chemical compositions and mechanical properties of these materials are shown in Table1 and Table2, respectively. As shown in Figure 1 the dumbbell type round bar specimen with 3mm at minimum diameter was used for ultrasonic fatigue tests. Ultrasonic fatigue tests were conducted by use of an ultrasonic fatigue testing machine (Shimadzu USF2000) at room temperature. Frequency was 20kHz and R (the ratio of minimum to maximum stress in the loading cycle) was -1.Testing environments were distilled water (equivalent to air) and 3% NaCl aqueous solution. Because of the low thermal conductivity in austenitic stainless steels ultrasonic fatigue tests must be carefully conducted to prevent heating in specimens during ultrasonic fatigue tests. In this study a compressed air, a solution circulation system and intermittent testing method were used in order to prevent heating during fatigue and corrosion fatigue tests. The compressed air was blown into to the center of the specimen during ultrasonic fatigue testing. Aqueous solution was circulated with a speed of 3l/min. The intermittent frequency was 110ms duty and 1100ms pause. Surfaces and fracture surfaces were observed by an optical microscope (Keyence VHX-100) and a scanning electron microscope (JEOL,JSMS5500S),respectively.

Table1 Chemical compositions of tested materials (mass%).

Material	C	Si	Mn	P	S	Ni	Cr	Mo	Nb	N
YUS270	0.08	0.46	0.46	0.02	0.01	18.9	19.9	6.16		
SUS304N2	0.004	0.81	0.082	0.029	0.001	7.55	19.45		0.09	0.02

Table2 Mechanical properties of tested materials.

Material	$\sigma_{0.2}$ (MPa)	Ultimate tensile strength (MPa)	Elongation(%)	Hardness (HB)
YUS270	365	743	44	175
SUS304N2	442	766	50	213

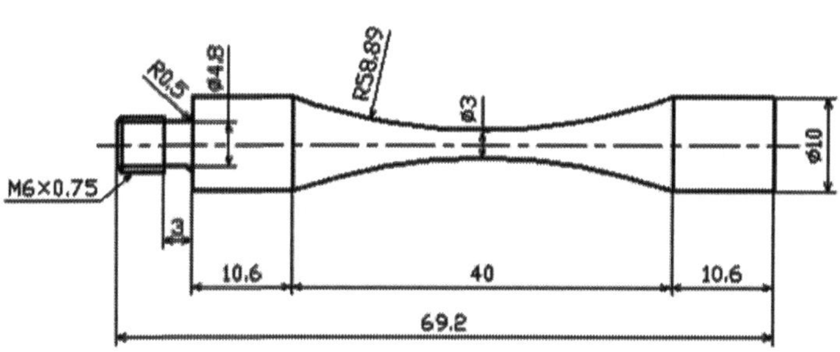

Figure 1: Fatigue test specimen. Unit, mm

Results and Discussion

Figure 2 shows S-N diagrams of YUS270 in air and in 3%NaCl aqueous solution. It is apparent that the effect of 3% NaCl aqueous solution was not observed at 10^9 cycles. However, the effect of 3% NaCl aqueous solution on fatigue life was observed at stress amplitude higher than 350 MPa. The same inclination was also observed on SUS304N2. The effect of 3% NaCl aqueous solution was not observed at 10^9 cycles and the effect of 3%NaCl aqueous solution on fatigue life was observed at stress amplitude higher than 340MPa as shown in Figure 3. The reduction of

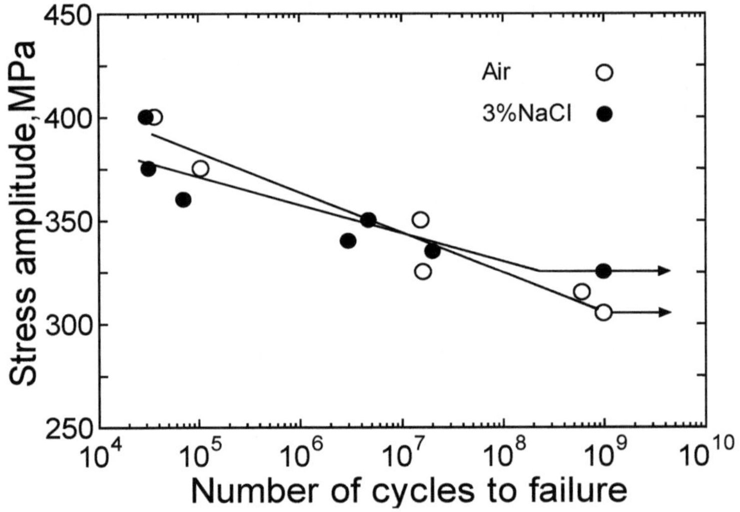

Figure 2. **S-N diagrams of YUS270 in air and in 3%NaCl aqueous solution.**

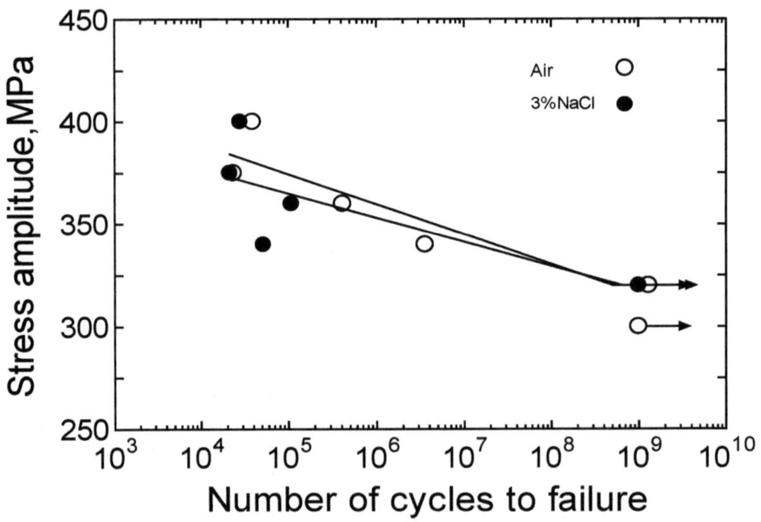

Figure3. **S-N diagrams of SUS304N2 in air and in 3%NaCl aqueous solution.**

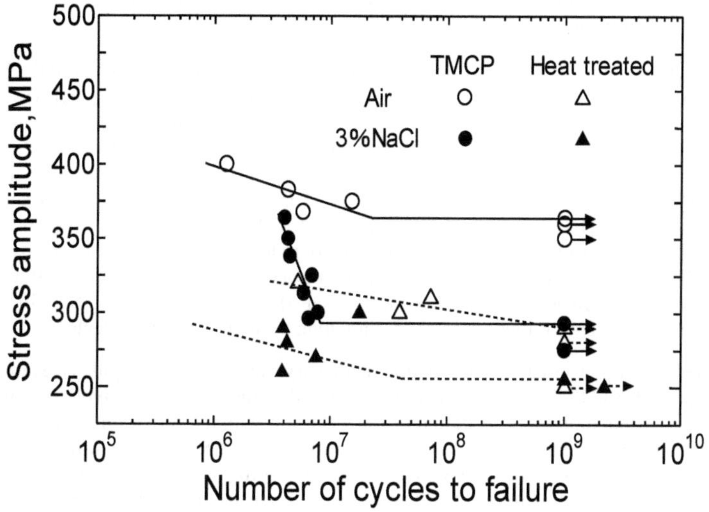

Figure 4. **S-N diagrams of 2.5%Mo containing austenitic stainless steel in air and in 3% NaCl aqueous solution [5].**

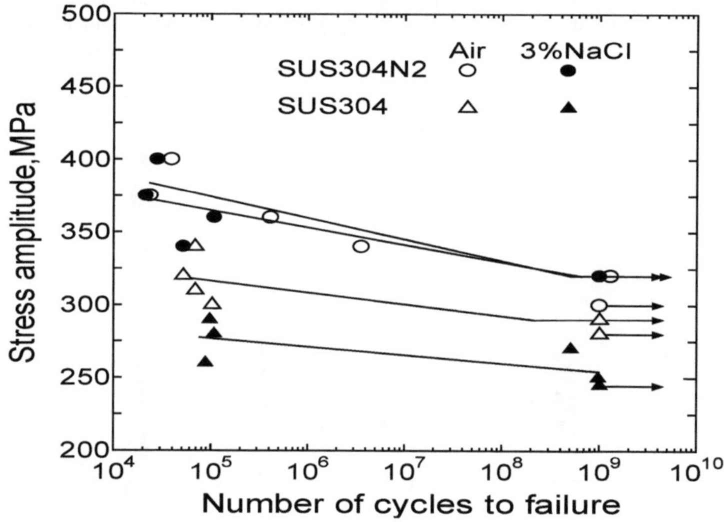

Figure 5. S-N diagrams of SUS304 N2 and SUS304 in air and in 3% NaCl aqueous solution.

corrosion fatigue strength at 10^9 cycles for solution treated steel(σ_B 745MPa) and TMCP made steel(σ_B 905MPa) of 2.5% Mo containing high strength austenitic stainless steel was 12.1% and 19.5 %, respectively as shown in Figure 4. The different reduction rate of giga-cycle corrosion fatigue strength in high strength austenitic stainless steels is depending on the corrosion resistance of austenitic stainless steel. Figure 5 shows the former reported S-N diagrams of SUS304[5] in comparison with those of SUS304N2 in air and 3%NaCl aqueous solution. The reduction of giga-cycle corrosion fatigue strength of SUS304 is 15.5%[5].It can be mentioned that the higher corrosion fatigue strength and no reduction of giga-cycle corrosion fatigue strength in SUS304N2 are attributed to high nitrogen content. Corrosion fatigue crack in association with corrosion pits were observed on specimen surface for SUS304, SUS316, solution treated and TMCP made 2.5% Mo containing austenitic stainless steels.

The corrosion fatigue crack in association with very small corrosion pits is exemplified for YUS270 and SUS304N2 as shown in Figure 6a) and Figure 7a) ,respectively. On the contrast fatigue crack in air looks like to be almost straight and without corrosion pit (Figurer 6b) and Figure7b)).The same phenomenon was also observed on NSSC250 as shown in Figure 8[5]. Therefore very small corrosion pit was observed on corrosion fatigue crack initiation area in fracture surface. Figure 9 shows the typical example of corrosion pit observed at crack initiation area for SUS304N2. Striation was predominantly observed on corrosion fatigue crack propagation area in 3%NaCl aqueous solution as shown in Figure10.These fracture surface morphologies were also observed for SUS304,SUS36 and 2.5% Mo containing austenitic stainless steel. The fracture surface morphologies in crack propagation area were quite different from that of the 12Cr martensitic stainless steel[5]. From these surface and fracture surface observation results it can be concluded that the reduction of corrosion fatigue strength of high strength austenitic stainless steels is due to the corrosion pit formation at corrosion fatigue crack initiation area. This fact has been confirmed in corrosion fatigue strength for 2.5% Mo containing austenitic stainless steel at conventional frequency[3],

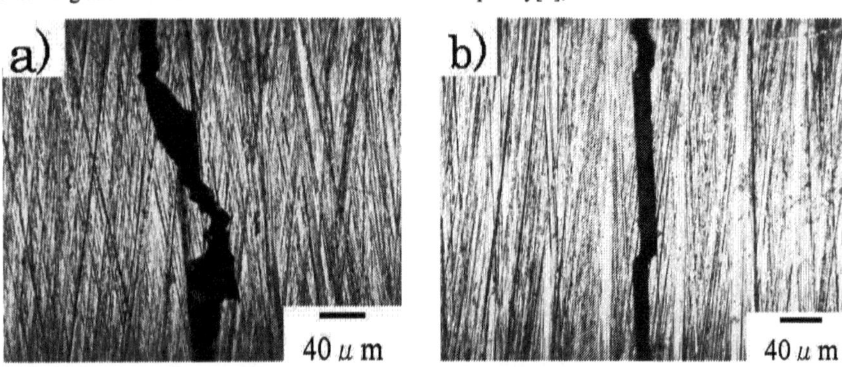

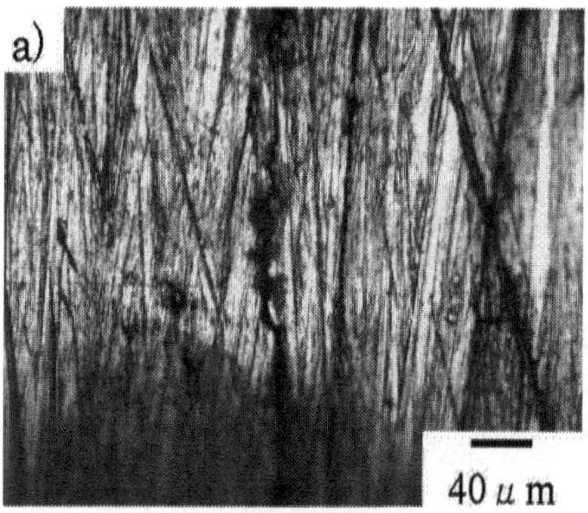

Figure 6. Fatigue crack observed on specimen surface.YUS270, 350MPa (a) 3% NaCl, 14.78×10^{6} cycles (b) Air, 1.55x10^{7} cycles

Figure 7. Fatigue crack observed on specimen surface.SUS304N2, 360-MPa
(a) 3% NaCl, 1.09x10^{5} cycles (b) Air, 4.15x10^{5} cycles

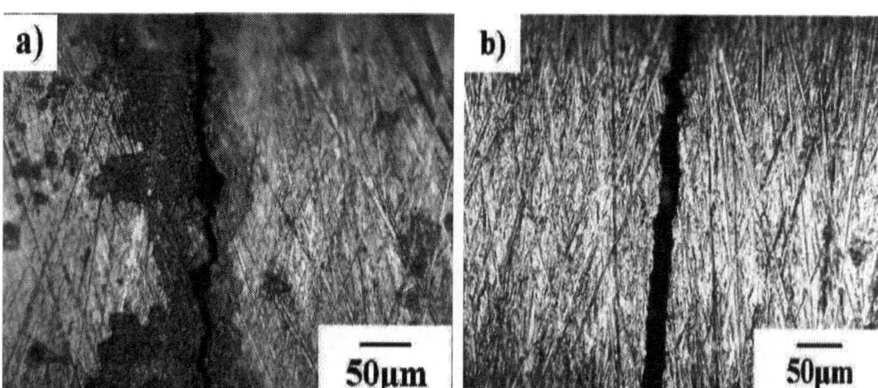

Figure 8. Fatigue crack observed on specimen surface 2.5% Mo containing austenitic stainless steel
(a) 3% NaCl aqueous solution, 260 MPa, 3.6x10^{8}cycles;
(b) Air, 320 MPa, 5.5x10^{6} cycles

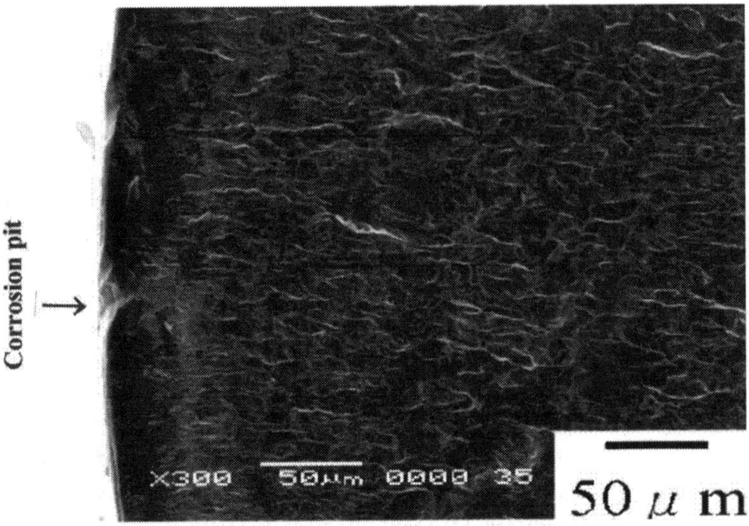

Figure 9. **Corrosion pit at initiation area of fracture surface. SUS304N2, 3% NaCl aqueous solution. 360MPa, 1.09×10^5 cycles**

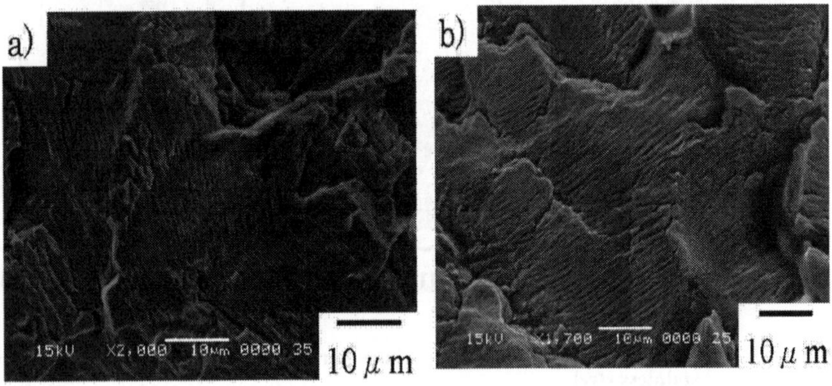

Figure 10. **Crack propagation area of fracture surface.**

3% NaCl aqueous solution. 375MPa, 1.5mm from initiation
(a) SUS304N2, 2.15 x 10^5cycles (b) YUS270, 3.21x10^4cycles

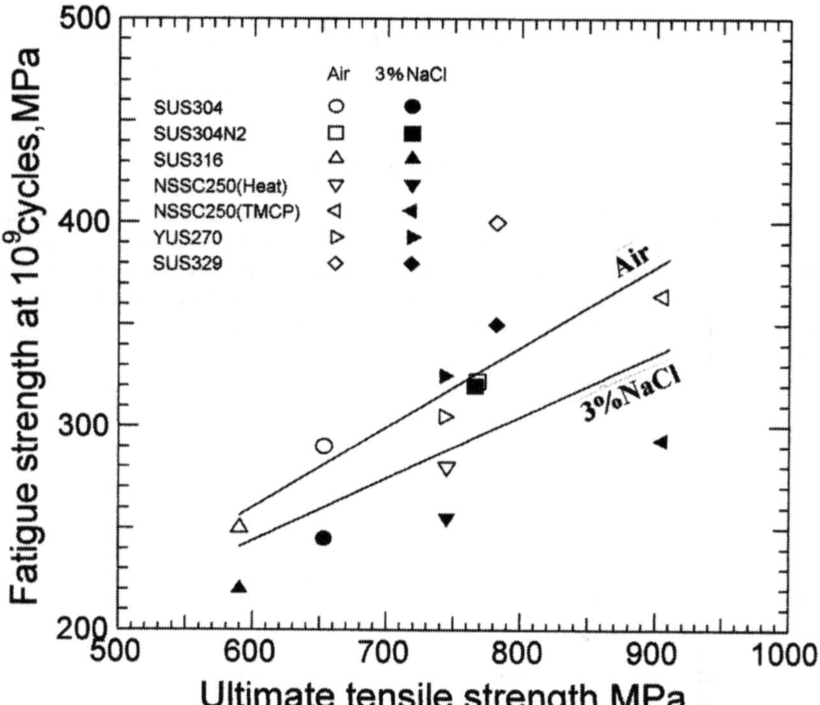

Figure 11. Relationship between ultimate tensile strength and giga-cycle fatigue strength of austenitic stainless steels in air and in 3%NaCl aqueous solution.

Fig.11 shows the giga-cycle fatigue strength of stainless steels in air and in 3%NaCl aqueous solution as a function of ultimate tensile strength. Giga-cycle fatigue strength in air and giga cycle corrosion fatigue strength in 3%NaCl aqueous solution on austenitic stainless steels such as SUS304,SUS304N2,SUS316,NSSC250 and YUS270 are summarized. The giga-cycle fatigue strength and giga-cycle corrosion fatigue strength in 3% NaCl aqueous solution of SUS329 is also plotted in this figure. The higher the ultimate tensile strength the higher the giga-cycle fatigue strength in air is. However, it can be clearly observed that the higher the ultimate tensile strength the higher the giga-cycle corrosion fatigue strength in 3%NaCl aqueous solution is. It is also observed that the effect of 3%NaCl aqueous solution is higher in the higher ultimate tensile strength. The giga-cycle fatigue strength in air showed the same inclination as conventional fatigue strength for carbon steel, while giga-cycle corrosion fatigue strength in 3% NaCl aqueous solution showed quite different inclination from conventional corrosion fatigue strength for carbon steel[7].This means that giga-cycle fatigue strength of austenitic stainless steels is not significantly influenced by 3%NaCl aqueous solution. The giga-cycle fatigue strength of SUS329 showed higher value than those of austenitic stainless steels. This means that corrosion resistance of duplex stainless steel in 3% NaCl aqueous solution seems to be different from that of austenitic stainless steels.

Concluding remarks

It can be concluded that the giga-cycle corrosion fatigue strength of high strength austenitic stainless steels is not significantly influenced by 3% NaCl aqueous solution. The corrosion fatigue strength of high strength austenitic stainless steels is governed by corrosion pit formed at corrosion fatigue crack initiation area. The higher the ultimate strength the higher the giga-cycle corrosion fatigue strength of austenitic stainless steels in 3%NaCl aqueous solution is. In order to evaluate corrosion fatigue life of high strength austenitic stainless steels further investigation on corrosion pit initiation mechanism of austenitic stainless steels is recommended to conduct.

References

1. R. Ebara, **Procedia Engineering**,2(2010)1297-1306.

2 K.Hirakawa and I. Kitaura, *The Sumitomo Search*,No.26(1981)136-151(in Japanese).

3. R.Ebara,K.Matsumoto,Y.Matsuda,Y.Mizuno and I.Nishimura,*Fatigue of Materials,Advances and Emergences in Understanding*,T.S.Srivastsan and M.Ashraf Imam edit. 381-392,2010,TMS,Wiely.

4. R.Ebara, **International Journal of Fatigue**, 28 (2006),1465-1470.

5. R.Ebara,Yamad,and A.Goto,Ultrasonic Fatigue,*Proc. of the First Intern. Conf. on Fatigue and Corrosion Fatigue up to Ultrasonic Frequencies*,Wells,J.M;Buck,O;Roth,L,D;Tien,J,K; edit.,The Metallurgical Society of AIME,349-364,1982.

6. R.Ebara,Y.Yamaguchi,D.Kanei,T.Ota and Y.Miyoshi,*Proc.of the fifth Intern. Conf.on Very High Cycle Fatigue*, C.Berger and H.-J.Christ (editors), pp. 275-280, 2011.

7. R.Ebara,**Trans JSME,SeriesA**, 59 (1993) No.557, pp1-11 (in Japanese).

Fatigue of Materials II Advances and Emergences in Understanding
Edited by: T.S. Srivatsan, M. Ashraf Imam, and R. Srinivasan
TMS (The Minerals, Metals & Materials Society), 2013

Influence of Microstructural Features on the Propagation of Microstructurally Short Fatigue Cracks in Structural Steels

M. Sharaf[1]; J. Lian[1]; N. Vajragupta[1]; S. Münstermann[1]; W. Bleck[1];
B. Schmaling[2]; A. Ma[2]; A. Hartmaier[2]

[1]Department of Ferrous Metallurgy, RWTH Aachen University, Intzestr. 1, Aachen, Germany
[2]Interdisciplinary Centre for Advanced Materials Simulation (ICAMS), Ruhr-Universität Bochum, Germany

Keywords:
Defects, fatigue tests, endurance strength, finite elements, short cracks

Abstract

Cyclically loaded structural steel components are usually designed to endure macroscopic stress amplitudes close to the material's endurance strength where microcracks initiate due to microstructural inhomogeneities and exhibit strong interactions with the various microstructural features in their neighborhood upon propagating. The current study presents a microstructural model with a capability to quantitatively describe the influence of microstructural features on the growth of cyclic cracks in the decisive, very early fatigue behavior stage. The FE model is based on the crystal plasticity theory and accounts for relative grain orientations. Both the extended finite element method (XFEM) and a coupled damage mechanics approach are used to describe crack opening behavior. The model is implemented to simulate real microcracking events produced in interrupted cyclic multiple-step tests under metallographic observation with temperature change measurements. Furthermore, the model is implemented on virtually created microstructures with altered grain sizes and orientations based on statistical EBSD analysis.

Introduction

Several modern high strength steels have been developed in the last three decades with the objective of achieving more sustainable use of natural resources and improved utilisation of mechanical properties [1,2]. Nevertheless, for most of these steel grades the endurance strength could not be improved as much as the yield strength, so that for many engineering applications, the safety requirements in cyclic loading conditions become the limiting factor of lightweight design [3,4]. Some successful efforts have been made to increase the endurance fatigue limit of high strength steels by further enhancing the material cleanliness by means of sophisticated remelting processes [5]. The precise estimation of the endurance strength of these high strength steels is quite difficult as their fatigue testing data always exhibit wide scatter bands. It can be debated that for these wide scatter bands, the less remaining particles are the main cause [2,5-7]. It has been shown previously that the high intrinsic fatigue limit of a high strength steel matrix can be achieved if the inclusion size is smaller than a critical value [8,9]. Slip bands or grain boundaries nucleated in a matrix can become fatigue crack initiation sites [3]. These and other findings point out to the need to quantify the critical size of inclusions to express a significant influence on the endurance fatigue limit. For this purpose, a quantitative model has to be developed which considers the influence of microstructure on the cyclic properties of the material.

It was shown in recent studies that micro cracks often nucleate after relatively few loading cycles but repeatedly come to a condition where they cannot overcome microstructural obstacles, e.g. grain boundaries [3,10-15]. This happens when the local stress is not high enough. The required model has therefore to consider necessary stereological parameters and crystal orientations within the surroundings of the micro crack. Innovative experimental methods which realise cyclic load variation are essential for decreasing the number of load cycles at micro crack initiation and thereby decreasing the time needed to simulate the cyclic test. Dengel and Harig [16] offered a load increasing testing technique with continuous temperature measurement. The micro plastic deformations which take place after few cycles cause local initiation of very short micro cracks. The heat energy released thereby can be determined by precise temperature measurement. The endurance strength corresponds to the stress amplitude level at which first micro plastic deformations, indicated by temperature rise, are observed. Eifler et al. [1718] considered various steels in load increasing tests with additional measurements of plastic strain and electrical resistance. This approach was carried out in the current study on notched bending specimens in order to induce the initiation of very short micro cracks and not to estimate endurance fatigue limit. A numerical model based on the achieved results is presented for the simulation of the whole cyclic micro cracking event, including crack initiation and propagation along a non-predefined path.

Experimental Investigation and Results

The material used in the experiments is a fine grained structural steel (S355 J2G3). It was delivered in the form of hot-rolled 15 mm thick sheets. The average grain size is 6.85 μm and the content of pearlite is 14.5 %.

The geometry of specimens used is presented in **Fig. 1**. All specimens were subjected to 4-point-bending tests. A 1 mm deep notch was machined at the middle of the specimen's length. The single notch guarantees crack initiation on a limited flat area on a ground, etched face of the specimen at the notch tip, where the highest tensile stresses are located.

In order to assess the stress level at which micro cracks appear, the endurance strength was estimated in fatigue tests at constant stress amplitude. The local stresses in the notch root were calculated in a simple finite element simulation and were ca. 30% higher than the nominal values. Although a relatively small number of specimens were investigated, the obtained results show that the endurance fatigue limit correlates with the stress amplitude of 150 – 160 MPa.

Load increasing tests for the initiation of micro cracks and study of their propagation were carried out with 7 bending specimens on a 40 kN servo-hydraulic testing machine at a frequency of 5 Hz and a stress ratio of 0.1. The propagation of the micro cracks was observed under light optical microscope (LOM) and scanning electron microscope (SEM) in defined intervals, where the tests were interrupted. The start stress amplitude was always 20 MPa. It was increased by 20 MPa every 9000 cycles until a nominal stress amplitude of 160 MPa was reached. The number of 9000 cycles assured enough time for the initiation of a micro crack before the next load step.

Further loading intervals were chosen individually for each specimen; as an example the procedure for specimen No. 7 is shown in **Table 1**.

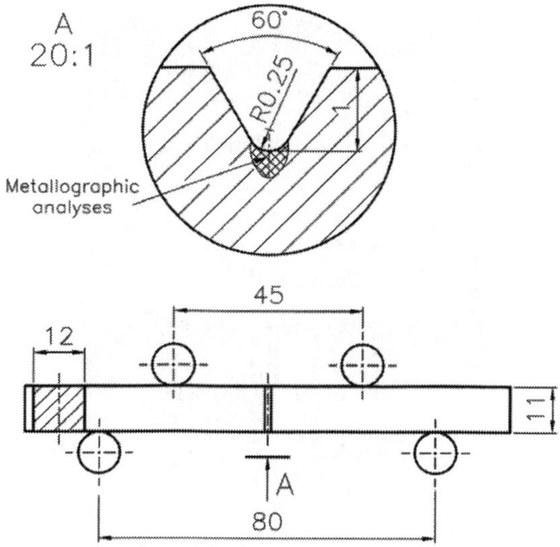

Fig. 1: Schematic diagram of the experimental set-up and geometry of used specimens

sample no. 7	stress amplitude	number of cycles per interval	number of cracks	length of crack #1	length of crack #2	length of crack #3
[-]	[MPa]	[-]	[-]	[µm]	[µm]	[µm]
Interval 01	20 - 160	9,000	0	0	0	0
Interval 02	160 and 170	9,000	3	15	5	3
Interval 03	160	9,000	3	15	5	3
Interval 04	170	9,000	5	15	10	8
Interval 05	170	9,000	5	18	10	8
Interval 06	170	9,000	5	18	10	8

Table 1: Testing procedure for specimen no. 7

The formed micro crack of a length of 18 µm was particularly well visualised for sample no. 7. The microstructure before testing and after the second interval is shown in **Fig. 2**. Analysis with an electron backscattered diffraction (EBSD) facility was carried out for this specimen in order to obtain the orientation of the grains in the neighbourhood of the crack. The existence of a grain boundary between two favourably disoriented ferrite grains in the neighbourhood of a micro pore near the notch tip was a crucial factor for setting the initial crack growth direction and supported its further propagation through two ferrite grains towards a pearlite island.

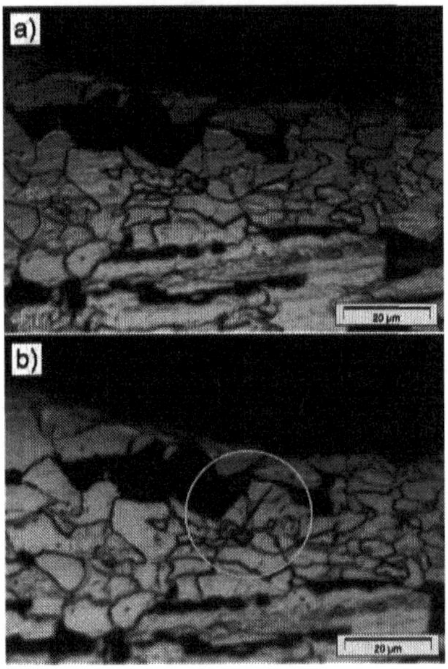

Fig. 2: Microstructure of specimen no. 7 before testing (a) and after second loading interval (b)

Numerical Investigation and Results

The bulk material was discretised by four-noded bilinear extended finite elements [19]. An alternative approach within the framework of the extended finite element method (XFEM) based on traction-separation cohesive behaviour was implemented in order to allow for modelling moving cracks. Remmers [20] and Song [21] used this approach to simulate multiple cracking of solids. The approach shows almost no mesh dependence, if the mesh is sufficiently fine [22]. Phantom nodes are superimposed on the original real nodes, such that every phantom node is constrained to its corresponding real node till fracture. Cracked elements split into two cohesive segments whose separation is governed by a cohesive law. Here, the cohesive surface model of Xu and Needleman [23, 24] was implemented. This cohesive model couples normal and tangential separations and can briefly be demonstrated by the following equation:

$$T_n = \frac{\sigma_{\max}\Delta_n}{\delta_n} e^{1-\frac{\Delta_n}{\delta_n}} \qquad (1)$$

where T_n is the normal traction across the cohesive surface, $\sigma_{\max}$ is the cohesive surface normal strength, Δ_n is the normal separation displacement and δ_n is the characteristic normal separation displacement at $\sigma_{\max}$.

Provided that the computation with the model is force-controlled, and assuming that the first loading cycle induces damage initiation in one or more elements, every further loading cycle will induce gradual stiffness degradation, which takes place following the governing cohesive law. A solution-dependent crack propagation path is realised as result of the reduction in load carrying capacity of cracked elements, which corresponds to similarly gradual stress and strain concentration rise in neighbouring elements, which therefore incrementally approach crack initiation.

The parameters of the cohesive law were fitted in an optimization study. Acceptable ranges of values for each of those parameters were thereby determined, which lead to the best possible convergence rate. The model was implemented to simulate the cracking event in specimen no. 7. A two dimensional finite element mesh was created based on the microstructure morphology in the neighbourhood of the crack. Grain orientation data obtained from the EBSD analysis, in terms of Euler angles, was used to rotate the elastic stiffness matrix of grain.

Cyclic force boundary conditions were applied to the nodes on the right edge of the model. Movement of the nodes on the left and top edges were constrained in x and y directions, respectively. As the cyclic crack advanced through the elements, the computation was allowed to run until deterioration of convergence (**Fig. 3**).

In the computation, the crack started at a similar position at the pore as detected in experiment. It advanced transcrystalline in a comparable growth direction parallel to the interface between the two favourably disoriented ferrite grains.

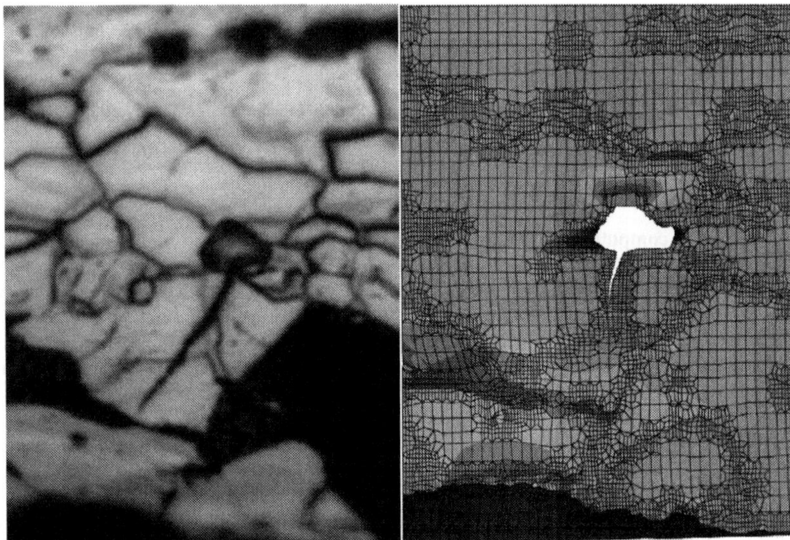

Fig. 3: Light optical micrograph of cracked microstructure after the 2nd loading interval and calculated crack path after 70 loading cycles, colours ranging from blue, denoting low maximum principal stress values, to red, denoting high principal stress values

Discussion

The model presented has shown its capability to describe the influence of microstructure on fatigue behaviour of short micro cracks. The mechanism-oriented computation was able to successfully simulate the crack initiation locus and its early propagation direction but still not at the propagation rate observed in the experiment. This was most probably because the model considered damage evolution but did not allow for microplastic deformations. Furthermore, the influence of grain boundary behaviour is not yet considered. To take the dissipative crystal slip behaviour of atoms into consideration, the authors plan to include a constitutive definition of crystal plasticity into the model.

The utilised experimental setup and the geometry of tested samples ensured the highest tensile stresses on the outer surface, so that the application of two dimensional elements in simulations could be assumed.

Based upon the findings presented, the metallographic analyses showed that all initiated micro cracks were detected only in ferrite grains. This observation agrees with literature research findings [25]. Occasionally, few micro cracks were created but always only one, which was not always the initial crack, grew further through the microstructure. Initially, very short cracks propagated across the ferrite grains. Upon further cyclic loading, they propagated along the grain boundaries. Noteworthy, any crack that overcame the first few grains was not able to stop even at significantly decreased stress amplitudes.

Within the scope of this research topic, the authors intend to use the model, in virtually created microstructure FE-meshes with random grain sizes and crystallographic orientations based on statistical data obtained by metallographic analysis, so that a method enabling variation of microstructural features is realised. Statistical microstructure data obtained by metallographic analysis will be input to the Voronoi algorithm [26] which generates artificial microstructures in a mathematically adaptive scheme that simulates grain evolution during solidification of metal melt. Thereby, the dependence of fatigue lifetime on microstructural parameters can be studied, when all the stereological parameters and crystallographic orientations of the ferritic-pearlitic steel are arbitrarily altered.

References

[1] Y. Murakamia, S. Kodamab, S. Konumac, Quantitative evaluation of effects of non-metallic inclusions on fatigue strength of high strength steels. I: Basic fatigue mechanism and evaluation of correlation between the fatigue fracture stress and the size and location of non-metallic inclusions, Int. J. Fatigue 11 (1989) 291-298

[2] J.M. Zhang, S.X. Li, Z.G. Yang. G.Y. Li, W.J. Hui, Y.Q. Weng, Influence of inclusion size on fatigue behavior of high strength steels in the gigacycle fatigue regime, Int. J. Fatigue 29 (2007) 765-771

[3] Y. Murakami, Metal Fatigue: Effects of small defects and non-metallic inclusions, Chapter 6, Elsevier (2002) 75-128

[4] M. Sander, Sicherheit und Betriebsfestigkeit von Maschinen und Anlagen, ISBN 978-3-540-77732-8, Springer-Verlag (2008)

[5] M.G. Hebsur, K.P. Abraham and Y.V.R.K. Prasad, Effect of electroslag refining on the fracture toughness and fatigue crack propagation rates in heat treated AISI 4340 steel, Eng. Frac. Mech. 13 (1980) 851-864

[6] P.R.V. Evans, N.B. Owen, B.E. Hopkins, The effect of purity on fatigue crack growth in a high-strength steel, Eng. Frac. Mech. 3 (1971) 463-468

[7] Y. Murakami, M. Endo, Effects of defects, inclusions and inhomogeneities on fatigue strength, Int. J. Fatigue 16 (1994) 163-182

[8] Y. Murakami, M. Endo, A geometrical parameter for the quantitative estimation of the effects of small defects on fatigue strength of metals, Trans. Jpn. SOC. Mech. Eng. A 49(438) (1983) 127-136

[9] Y. Murakami, S. Fukuda, T. Endo, Effect of micro-hole on fatigue strength (lst Report, Effect of micro-hole (Dia.: 40, 50, 80, 100 and 200 pm) on the Fatigue Strength of 0.13% and 0.46% Carbon Steels), Trans. Jpn. Soc. Mech. Eng. Ser. I, 44(388) (1978) 4003-4013

[10] Y. Murakami, T. Endo, The effects of small defects on the fatigue strength of hard steels, In: E Sherrat and J.B. Sturgeon (Eds.): Materials, Experimentation and Design in Fatigue, Proc. Fatigue (1981) 431-440

[11] Y. Murakami, S. Kodama, S. Konuma, Quantitative evaluation of effects of non-metallic inclusions on fatigue strength of high strength steel, Trans. Jpn. SOC. Mech. Eng. A 54(500) (1988) 688-696

[12] Y. Murakami, M. Shimizu, Effects of non-metallic inclusions, small defects and small cracks on fatigue strength of metals, Trans. Jpn. SOC. Mech. Eng. A 54(499) (1988) 413-425
[13] Y. Murakami, H. Usuki, Prediction of fatigue strength of high-strength steels based on statistical evaluation of inclusion size, Trans. Jpn. Soc. Mech. Eng. A 55(510) (1989) 213-221
[14] Y. Murakami, Quantitative evaluation of effects of defects and non-metallic inclusions of fatigue strength of metals, Tetsu To Hagane 75(8) (1989) 1267-1277
[15] Y. Murakami, Fundamental aspects of fatigue threshold of metals containing small defects, small cracks and non-metallic inclusions - A unified quantitative evaluation and tts application, In: P. Lukas and J. Polak (Eds), Basic mechanisms in fatigue of metals, ACADEMIA (1988) 343-350
[16] D. Dengel, H. Harig, Estimation of the fatigue limit by progressively-increasing load tests, Fatigue Eng. Mater. Struct. 3 (1980) 113-128.
[17] P. Starke, F. Walter, D. Eifler, Fatigue assessment and fatigue life calculation of quenched and tempered SAE 4140 steel based on stress-strain hysteresis, temperature and electrical resistance measurements, Fat. Fract. Eng. Mater. Struct. 30 (2007) 1044-1051.
[18] F. Walther, D. Eifler, Fatigue life calculation of metallic materials, Key Eng. Mat. 345–346 (2007) 1337–1340.
[19] T. Belytschko, T. Black, Elastic crack growth in finite elements with minimal remeshing, Int. J. Numer. Meth. Eng. 45 (1999) 601-620
[20] J. J.C. Remmers, R. de Borst, A. Needleman, The simulation of dynamic crack propagation using the cohesive segments method, Journal of the Mechanics and Physics of Solids 56 (2008) 70–92
[21] J.-H. Song, P. M. A. Areias, T. Belytschko, A method for dynamic crack and shear band propagation with phantom nodes, Int. J. Numer. Meth. Eng. 67 (2006) 868–893
[22] Dassault Systèmes Simulia Corp., Abaqus Analysis User's Manual, Abaqus 6.11 Online Documentation (2011)
[23] X.-P. Xu, A. Needleman, Numerical Simulations of Fast Crack Growth in Brittle Solids, J. Mech. Phys. Solids 42 (1994) 1397-1434
[24] S. Brinckmann, On the role of dislocations in fatigue crack initiation, doctoral dissertation, Rijksuniversiteit Groningen, ISBN 90-367-2237-3 (2005)
[25] D. Munz, Ermüdungsverhalten metallischer Werkstoffe, Deutsche Gesellschaft für Metallkunde (1985) 73-105
[26] G.F. Voronoi, Nouvelles applications des paramètres continuis à la théorie des formes quadratiques, J. Reine Angew. Math. 134 (1908) 198-287

AUTHOR INDEX
Fatigue of Materials II: Advances and Emergences in Understanding

S

T

V

W

X

Y

Z

SUBJECT INDEX
Fatigue of Materials II: Advances and Emergences in Understanding